Time, Frequency, and Phase: Keys to Unlocking Multistatic Radar Performance

Sachin

Contents

Chapter 1

Introduction

The field of bistatic radar and more specifically multistatic and networked radar has received much interest in the recent years. Bistatic radar has been around for nearly 80 years. However, relatively recent technological advances in data transfer rates, computing and synchronisation are now making more complex bistatic systems practically possible. [1, 2] Much of the advantage of a multistatic radar rests on the various possible layers of cooperation among the radar nodes. The multiple view angles provide better spatial diversity and increase the amount of extracted target information. The overall sensitivity is increased, and there is a higher probability of viewing a stealthy target from a high radar cross section (RCS) angle. Simultaneously, the probability of obscuration is reduced, and target triangulation is possible. Passive receive-only nodes can be entirely covert and harder to detect and jam. The increased redundancy is also thought to make it less vulnerable to failure or attack. [3, 4] Passive receive-only nodes also have the potential to be cheap and simple [2]. More recently, intelligent cognitive radars were described where highly cooperative systems adapt in response to the environment [5]. It is clear that a multistatic system has many advantages, but it is dependent on the system geometry, the modes of operation, the level of node cooperation, and on the space-time coherence [4]. The space-time synchronisation required for node cooperation and data fusion adds significant complexity in comparison to a monostatic system. Especially, for phase-coherent systems. [1, 2, 6]

This text is primarily concerned with the time, frequency and phase synchronisation of bistatic or multistatic networked radar. Moreover, the exact pulse transmission time must be known network-wide to perform range measurements. Also, the carrier frequencies and phases of the nodes must be synchronised for Doppler estimation and coherent integration, respectively. [7–9] Additionally, the phase noise requirements of the reference oscillator in bistatic radar is much more stringent than in monostatic systems [10]. It is worth noting that there is fairly little published on the time, frequency and phase synchronisation of bistatic radar in the open literature. Most such publications focus on the synchronisation of air- or spaceborne bistatic synthetic aperture array (BSAR) systems using direct radio frequency (RF) or optical links across relatively short distances. One example is the TanDEM-X interferometer [11]. A few texts address the synchronisation requirements of BSAR as well as motion compensation techniques to correct clock drifts. However, discussions relating to bistatic synchronisation requirements of pulsed-Doppler systems from a systems perspective are often superficial. Some publications reported on the synchronisation of real ground-based bistatic radars. Most of these bistatic systems are experimental and used cables (copper or optic fibre) [12–14], RF links [11, 14], the radar transmissions (breakthrough, clutter reflections, or tropospheric scatter) [7, 9, 15–17], or global positioning system (GPS) [17] for synchronisation. However, the achieved bistatic time and frequency synchronisation are often merely quoted as single numbers. Very few articles report on the bistatic phase performance of ground-based systems. There is also little reported on the practical implementation of bistatic radar synchronisation.

Global navigation satellite system (GNSS) offers a low-cost and convenient mode of time transfer and is uniquely suited to network radar synchronisation. GNSS time transfer (specifically GPS) has become increasingly popular since it became available in 1992 [18]. It is always available free of charge nearly anywhere on earth and has become the primary method of time transfer [19]. The only infrastructure required is a GNSS timing receiver. It does not require line-of-sight (LOS) between the radar nodes, it can be autonomous, and it simultaneously provides an accurate means to determine the spatial coordinates of each node. GNSS receivers are passive and thus not at risk of enemy detection. Thus, GNSS offers a solution to the challenge associated with the time and synchronisation of bistatic radar. [1, 20] Perhaps more so in civilian applications where there is a lower risk of deliberate jamming or spoofing. Today there are various GNSS systems such as the Russian GLONASS, the European Galileo, and others. Multiple systems offer redundancy, and future systems will undoubtedly offer improved performance. [21] However, this text focusses on the US Department of Defense (DoD) owned GPS system.

Low-cost commercial GPS timing receivers deliver synchronisation in the form of a somewhat inconvenient 1 Hz time reference. This pulse per second (PPS) signal has long-term frequency stability traceable[1] to Coordinated Universal Time (UTC) but has high short-term jitter caused by atmospheric noise. Most timing applications, including bistatic radar, require a low phase noise 10 MHz reference with excellent long-term stability. In such cases, it is customary to GPS-steer a stable local oscillator (STALO) to correct its drift and offset continuously. Such a GPS-calibrated STALO is known as a GPS-disciplined oscillator (GPSDO).[22, 23] GPSDOs have been in use even before GPS became officially available [24]. As of May 2000, the RMS time jitter produced by civilian GPS receivers improved by a factor of five [25]. Moreover, the US DoD switched off the deliberate dithering, or selective availability (S/A), of the L1 coarse-acquisition (C/A) GPS carrier, and civilians now have access to its full accuracy [26]. Cheaper quartz OCXOs that are orders of magnitudes less stable can now deliver levels of synchronisation previously only possible with expensive Rubidium (Rb) atomic sources [25]. Thus, atomic-level frequency stability became available to the low-cost market when S/A was switched off. Further, low-cost one-way GPS timing receivers that are available today offer sub-20 ns time synchronisation and a frequency uncertainty relative to UTC of better than 4×10^{-12} after only 1000 seconds of averaging [19].

Wurman et al. [17] pioneered networked radar GPS-synchronisation in 1993 shortly after GPS became available. Their 3 GHz dual-Doppler networked radar prototype consisted of a single transmitter and multiple passive receivers synchronised using custom GPS-steered OCXOs. They used both the direct breakthrough[2] and GPS for synchronisation. Their system achieved time and frequency synchronisation of 1 μs and 1×10^{-9}, respectively, across tens of kilometres. However, they did not report on the phase synchronisation. It was estimated that the cost of one passive node was less than 3% of the cost of one transmitting node. Further, the designers reported that the additional costs associated with the multistatic system such as synchronisation, communication, data transfer and data integration are significantly less than the cost of a single transmitting node. Thus, Wurman et al. [17] demonstrated nearly 25 years ago that quartz-based GPS time transfer is a well suited and

---

[1] In this text, traceability means that a measured result is related to a reference through a chain of calibrations that are unbroken and where each step contributes to the total uncertainty.

[2] Direct breakthrough refers to the received signal that is caused by a direct transmission across the bistatic baseline through the transmitter-receiver sidelobes when there is line-of-sight visibility.

affordable alternative to synchronise certain types of multistatic radar. However, modern low-cost quartz GPSDOs are now orders of magnitude more stable making it perhaps even an attractive option for phase coherent systems. Most GPSDOs are self-contained and autonomous making them ideal for synchronising passive receivers without negating their advantage of being cheap and simple. Nevertheless, not many modern examples of GPSDO synchronised bistatic or multistatic radar could be found. There are no detailed synchronisation data published for GPSDO synchronised bistatic radar in the open literature. The phase synchronisation performance versus time elapsed is of particular importance for phase coherent multistatic systems.

The synchronisation of experimental systems using LOS breakthrough, clutter reflections and even tropospheric scatter, has been demonstrated in the literature since the early 1980's. [7, 9, 15, 16] The antenna sidelobe breakthrough was also used in combination with GPS to time and frequency synchronise multiple radar nodes. [17] Inggs[1] et al.[27] first demonstrated the possible efficacy of LOS phase compensation using GPS synchronised NetRAD data. However, it remains of interest to quantify the effectiveness of breakthrough phase compensation in improving the bistatic phase coherence of low-cost GPS synchronised radars. If effective, it has the potential to achieve high phase coherence at low cost.

Little is published on the implementation of GPS-synchronisation in bistatic radar [28]. However, there are some practical difficulties with the low-cost GPS-synchronisation of portable networked radar. Quartz oscillators often only reach optimal stability after days or months of ageing. Also, the initial GPSDO phase lock acquisition may require tens of minutes to a few hours depending on the STALO stability. Most laboratory GPSDO installations are static with precisely surveyed antenna coordinates, ideally placed antennas, and backup power supplies. Such GPSDOs rarely loses power or phase-lock and oscillator warm-up and long phase-locked loop (PLL) lock-in times are usually not a problem. However, a portable networked radar such as NetRAD [12, 13, 29] may require regular geometry changes each time requiring phase-lock re-acquisition. Phase lock may also be lost due to an unexpected loss of power, accidental GPS antenna blockage, or accidental or deliberate GPS signal jamming. Further, most low-cost commercial GPSDOs offer a 10 MHz output which is phase and frequency synchronised to GPS. However, such

---

[1]Special mention to my colleague W.A. Al-Ashwal who first used the breakthrough phase to mitigate GPSDO phase dynamics.

GPSDOs rarely offer time synchronisation better than that of the raw GPS PPS output. Moreover, a time synchronisation to UTC of better than 100 ns is often specified and limits the bistatic range accuracy to 30 metres. Baojian, Dehai, and Dazhi [30] took a novel two-step approach to GPSDO phase-locking. First, they implemented an frequency-locked loop (FLL) to frequency-lock the GPSDO. Then, they used a phase shifter to adjust independently the GPSDO phase to achieve phase-lock. This technique may improve both the lock-in time and timing performance. However, they reported little on the efficacy of this technique. There remains a large interest in cheap and simple quick-locking time synchronisation techniques. Especially, in techniques that are implementable on existing GPSDO hardware. Further, GPSDOs are often proprietary with little detail available about their inner workings [22]. Hence, it is of interest to have an open and reconfigurable development platform from where to conduct GPSDO research.

Lombardi, Novick, and Zhang [31] found that commercial GPSDOs may have very different phase-locked behaviours and that their specified performances may only be an estimate of their actual performance. Hence, commercial GPSDOs require extensive calibration and verification before integration within a networked radar. High-performance synchronisation equipment should also be re-calibrated often. [23] Most laboratories do not have access to specialist equipment to perform GPSDO calibrations. Hence, it is of interest to identify cheap and simple GPSDO calibration techniques.

The remainder of this text will evaluate the feasibility of synchronising a networked pulsed-Doppler radar using low-cost quartz GPSDOs. More specifically, NetRAD will be synchronised using custom designed and fully calibrated GPSDOs. NetRAD is a coherent tri-node pulsed-Doppler network radar. The carrier frequency is at S-band (2.4 GHz) with a bandwidth of 50 MHz. NetRAD has been under development at University College London (UCL) since the early 2000's and was built using low-cost commercial-and-off-the-shelf (COTS) hardware on a budget of £4000. [12, 13, 29] Moreover, NetRAD is converted from a copper cable synchronised system to a fully wireless GPS synchronised networked radar. The achieved GPS synchronised bistatic time, phase, frequency, and Doppler performance will be reported. LOS phase compensation will be applied to the GPS synchronised radar data in an attempt to quantify the improvement in the bistatic phase coherence. Three low-cost quartz GPSDOs will be custom developed to synchronise NetRAD. These open-source GPSDOs will function as research tools to investigate network radar synchronisation. Moreover, they will mainly attempt to solve three practical difficulties. Firstly, they should be capable of providing network-wide time synchronisation. This time synchronisation should have low jitter and must be in phase with the 10 MHz GPSDO output. Secondly, the GPSDO feedback control loop should lock quick enough to allow the practical use of long PLL time constants in the field. Thirdly, the GPSDO software environment should allow the rapid development of sophisticated control loops in a high-level programming environment for testing directly on the GPSDO hardware.

The rest of this chapter presents the research hypothesis along with the relevant research questions. Then, the key contributions resulting from this research are listed. Finally, follows a **book** outline with a brief description of each chapter.

1.1 Hypothesis

Low-cost quartz GPSDOs, based on one-way GPS time transfer, are suitable to synchronise coherent, pulsed-Doppler, networked, radars with carrier frequencies of a few gigahertz and moderate bandwidths of tens of megahertz across short baselines of a few kilometres. LOS phase compensation can enhance the bistatic phase coherence of such a GPSDO synchronised bistatic radar to near monostatic performance levels.

Research Question I

What is the achieved short baseline NetRAD performance when synchronised with low-cost quartz GPSDOs?
Quartz GPSDOs are an affordable and relatively simple solution to synchronising multiple low-cost passive receivers. They can be autonomous and require minimal radar node cooperation. Nonetheless, there exist very few examples of bistatic pulsed-Doppler radar synchronised using quartz GPSDOs in the open literature. Moreover, there are no detailed reports on the achieved performance of such radars. This text will measure the timing, phase, frequency, and Doppler performance of NetRAD while synchronised using calibrated low-cost quartz GPSDOs.

Research Question II

How effective is LOS phase compensation in improving the performance of the GPSDO synchronised NetRAD?

Radar transmissions have been used since the 1980's to synchronise bistatic radar. [7, 9, 15, 16] There are also limited reports, where the antenna sidelobe breakthrough has been used in combination with GPS to time and frequency synchronise multiple passive bistatic receivers [17]. However, LOS phase compensation appears to be a viable and affordable method to enhance the phase coherence of a GPS synchronised bistatic radar [27]. In this text, LOS phase compensation will be applied to low-cost quartz GPSDO synchronised bistatic data with baselines of a few kilometres. Then, the LOS phase compensated phase, frequency, and Doppler performance is compared to that of low-cost quartz GPSDO synchronisation.

Research Question III

How is GPSDO synchronisation implemented in a networked radar?

Little is published on the implementation of GPS-synchronisation in bistatic radar. [28] Further, commercial GPSDO designs are usually proprietary. Their tracking loops may behave unexpectedly, and the PPS time synchronisation outputs are often out of phase with its 10 MHz frequency output. In some cases, only the high-jitter GPS PPS signal is available for time synchronisation. [31] Static laboratory GPSDOs rarely lose phase-lock and long PLL lock-in times are often not a problem. However, portable radar nodes are more likely to lose phase-lock due to a loss of power, an antenna blockage or enemy jamming. Further, a portable network radar such as NetRAD must be capable of rapid geometry changes. Thus, portable networked radars require quick-locking GPSDOs to allow rapid geometry chances and minimise the downtime caused by an accidental loss of phase lock. In this text, three low-cost quartz GPSDOs will be developed to synchronise NetRAD. These GPSDOs should be open-source, well-understood, calibrated, and suitable to study network radar synchronisation. They should be capable of establishing a network-wide time synchronisation that is phase synchronous with their 10 MHz frequency outputs. These GPSDOs should also be quick-locking to make the use of long PLL time constants practically possible.

Research Question IV

How is the bistatic STALO performance specified to meet a specific radar performance?

Most publications on bistatic synchronisation focus on air- or spaceborne BSAR systems using direct RF or optical links across relatively short distances. However, discussions relating to bistatic synchronisation requirements of pulsed-Doppler systems from a systems perspective are often superficial. This text will build on the existing literature [3, 7–10, 20, 32] on bistatic synchronisation to better specify the STALO synchronisation based on the required radar performance. More specifically, to estimate the required GPSDO time, frequency, and phase synchronisation to meet the range, Doppler and phase performance of NetRAD.

1.2 Novel Contributions

1. The NetRAD coherent pulsed-Doppler networked radar was GPSDO synchronised across baselines of up to 2.3 km using low-cost quartz GPSDOs. These GPSDOs were custom developed at University of Cape Town (UCT) and well calibrated. The bistatic time, phase, frequency, Doppler phase noise, and Doppler performance were recorded using both the direct sidelobe breakthrough and the reflections off of a large and static target (the Roman Rock lighthouse) at a monostatic range of 1.9 km. These results are thought to be unique and similar results have not yet been published elsewhere. (Chapter 6)

2. LOS phase compensation was applied to the GPSDO synchronised bistatic data across baselines of up to 2.3 km. The monostatic, bistatic and LOS phase compensated bistatic data are compared.The phase, frequency, Doppler phase noise, and range-Doppler performance of the monostatic, bistatic, and LOS phase compensated bistatic cases were compared. This range-Doppler data includes that of the static Roman Rock lighthouse at a monostatic range of 1.9 km. However, it also includes a recording of a speedboat doing circles at a bistatic range of 3.7 km. LOS phase compensation used in combination with low-cost quartz GPSDOs proved to be an affordable solution to high-performance short baseline bistatic systems. Moreover, LOS phase compensation restored the bistatic phase, frequency, phase noise, and Doppler

performance to near monostatic levels. These results are thought to be unique and similar results have not yet been published elsewhere. (Chapter 6)

3. Three unique and low-cost quartz-based GPSDOs were designed, manufactured, and calibrated. These GPSDOs were designed as research tools to study networked radar synchronisation. The design is open, well understood, versatile, reconfigurable, and the user has real-time access to all the subsystems. These GPSDOs include a novel time synchronisation mechanism, a novel quick-locking adaptive PLL, and a unique software environment allowing the rapid development and testing of new control algorithms. (Chapter 4)

4. The UCT GPSDO has a novel time synchronisation mechanism, called the epoch pulse (EP). This EP module produces an accurate GPS synchronised pulse at an arbitrary user-selected future time. Moreover, it offers the best estimate of true UTC at any given moment, and it is up to an order of magnitude more accurate than the raw PPS output of a low-cost GPS timing receiver. The EP is also phase synchronous to the 10 MHz GPSDO output. The EP was routinely used to establish an accurate network-wide time epoch during the NetRAD experiments. (Section 4.2.4)

5. The UCT GPSDO includes a novel PC-based development environment where sophisticated PLL filters can be rapidly developed in a high-level programming language for verification directly on the GPSDO hardware. This environment was used to implement outlier removal, phase locked detection, and ultimately a quick-locking adaptive PLL. (Section 4.3.1)

6. A new quick-locking adaptive PLL was designed and demonstrated. This PLL achieves phase lock up to four times faster than conventional PLLs. It does this by first detecting a state of phase-lock and then changing its bandwidth in real-time without altering the PLL damping. Today, this algorithm is implemented successfully in the newer NeXtRAD system. [33]. (Section 4.3.4)

7. A method to specify the synchronisation requirement based on the desired bistatic radar performance was developed. It built on the existing literature and derived the effect of imperfect bistatic transmitter-receiver (Tx-Rx) synchronisation on pulsed-Doppler radar performance from basic principles. Moreover, the sampled intermediate frequency (IF) signal was expressed as a function of

Tx-Rx synchronisation, radar pulse number, ADC sample number, and arbitrary future time. A set of plots are provided allowing a designer to determine the required level of synchronisation for a specific application rapidly. It was recognised that the Tx-Rx error could often be estimated as a constant frequency offset (linear phase ramp) over typical integration periods of a few seconds or less for STALOs and GPSDOs. Hence, bistatic pulse integration of a static target is equivalent to that of a target moving at a constant radial velocity. The range error was specified as an initial time offset and continuous time drift. The initial frequency offset is based on the desired velocity accuracy, and the allowed frequency drift is based on the desired Doppler resolution. The phase drift is expressed as an allowable time drift using oscillator hold-over parameters. This time drift simplifies to a frequency offset when the phase slope is linear over the integration period. (Chapter 3)

1.3 book Outline

Chapter 1 (this chapter), serves as an introduction to this book. It provides some background and defines the primary hypothesis and associated research ques-tions. The novel contributions produced by this work are also listed. Finally, it provides a concise chapter by chapter overview of this text.

Chapter 2, reviews the literature related to the time, frequency, and phase synchronisation of bistatic, multistatic, and networked radar systems. Bistatic systems may offer various advantages over that of monostatic radar, but implementing the required precise time, phase and frequency synchronisation remain one of the significant c hallenges. T he r elationship b etween b istatic t ime, f requency, a nd phase synchronisation and bistatic STALO specification i s s tudied. H ereafter, alternative synchronisation methods are compared. These include synchronisation via RF and optic fibre c ables, a cross R F o r o ptical f ree-space l inks, a nd u sing t he direct or reflected r adar e missions. H owever, G PS t ime t ransfer i s i dentified as uniquely suited to portable bistatic applications. Therefore, the various methods of GPS time transfer, their implementation and performances are reviewed in-depth. The focus is on low-cost one-way GPS time transfer. Then, follows an overview of GPSDO technology. Various GPSDO architectures, feedback control strategies, time synchronisation techniques, and performance results are summarised from the literature.

Finally, the few published examples of GPS synchronised bistatic (or multistatic) are considered.

Chapter 3, builds on the existing literature [3, 7–10, 20, 32] on bistatic synchronisation to better specify the STALO synchronisation based on the required radar performance. More specifically, to estimate the required GPSDO time, frequency, and phase synchronisation to meet the range, Doppler and phase performance of NetRAD. It models an imperfectly synchronised bistatic Tx-Rx pair to predict the effect of imperfect synchronisation on an otherwise ideal bistatic system. This model is based on the NetRAD system and used to directly relate oscillator performance to the bistatic range, Doppler and phase errors. The synchronisation requirements of bistatic non-coherent and coherent pulse integration are also derived. This model and error expressions are then used to define a set of synchronisation requirements for NetRAD based on the desired radar performance. Moreover, it is predicted that low-cost quartz GPSDOs are adequate to synchronise NetRAD. However, time synchronisation using one-way GPS time transfer will likely limit the range accuracy to less than twice its 3 m range resolution and not a fraction thereof. Also, a bistatic system with independent STALOs has a much more stringent close-in phase noise requirement than a monostatic system if a similar subclutter visibility (SCV) is required. Hence, the use of sidelobe breakthrough is proposed to improve the timing, and the use of LOS phase compensation to improve the bistatic phase performance cost-effectively across short baselines.

In Chapter 4, a set of three identical low-cost quartz GPSDOs are designed and built [34, 35]. These GPSDOs are calibrated using the testbench designed in Appendix C and later used to synchronise the three NetRAD nodes in Chapter 6. They were explicitly developed as research tools to investigate network radar synchronisation. They solved three practical difficulties. Firstly, they are capable of network-wide time synchronisation that is up to an order of magnitude more accurate than what is possible with the raw GPS PPS signals. Secondly, a quick-locking adaptive PLL, able to phase-lock up to four times quicker than a conventional PLL, is implemented. Thirdly, the graphical user interface (GUI) includes a development environment where sophisticated control loops can be rapidly developed on a PC for verification directly on the GPSDO hardware. This chapter discusses the GPSDO hardware, firmware and GUI design.

Chapter 5, calibrates the synchronisation performance of the UCT GPSDO. The relative time (phase), frequency, frequency stability and phase noise performance is measured at a zero baseline. The purpose is to calibrate the baseline performance and to verify correct GPSDO operation. It is shown how time-domain clock characterisation can be used to diagnose GPSDO behaviour and verify its performance. Both the hold-over and phase-locked performance for the case of multi-channel one-way GPS time transfer is calibrated. These zero-baseline measurements represent the best possible performance under comparatively ideal circumstances. The results are used to predict bistatic radar performance which is later compared to the real bistatic radar performance recorded in Chapter 6. It is expected that the zero-baseline results will deteriorate with increasing baselines.

In Chapter 6, NetRAD is synchronised using the UCT GPSDOs. A tristatic experiment is set up in Simon's Bay, Cape Town reaching baselines of up to 2.3 km. The achieved bistatic synchronisation is compared to the zero-baseline calibrations of Chapter 5, and it is assessed how well the GPS synchronised NetRAD meets the performance requirements defined in Section 3.6. Moreover, the sidelobe breakthrough and strong reflections from the Roman Rock lighthouse are used to assess the achieved time, phase, and frequency synchronisation. LOS phase compensation (see Section 3.3) is applied to the bistatic data, and the phase, frequency, and Doppler performance of the monostatic, bistatic and LOS phase compensated Roman Rock reflections are compared. Finally, bistatic pulse integration in the presence of a constant frequency offset (see Section 3.4) is studied using real bistatic data.

Chapter 7, concludes this text by applying the main findings to the principal hypothesis and associated research questions. Then, follows a broader chapter by chapter summary listing the novel contributions while discussing the validity and limitations of the research design, and future research.

Appendix A, reviews some basic concepts of frequency stability necessary to develop the model in Chapter 3, as well as, to measure and characterise the GPSDO performance in Chapter 5.

Appendix B, contains the datasheet of the Oscilloquartz 8788 oven-controlled crystal oscillator (OCXO). Additionally, the phase noise, temperature stability, and frequency drift as measured by Oscilloquartz are also included.

Appendix C, describes the design and calibration of the low-cost UCT dual-mixer time difference (DMTD) system. This UCT DMTD is used in Chapter 5 to calibrate the UCT GSPDOs and other laboratory standards. It is demonstrated that high-resolution frequency stability measurements can be done relatively cheaply with careful design. Valuable insight is gained into frequency stability design.

Chapter 2

Background

This chapter reviews the literature related to the time, frequency, and phase synchronisation of bistatic, multistatic, and networked radar systems. Such radar systems may offer various advantages over that of monostatic radar. However, the implementation of precise time, phase and frequency synchronisation remain one of the significant challenges. Hence, it is studied how the bistatic time, frequency, and phase synchronisation relate to bistatic STALO specification. Hereafter, synchronisation alternatives are considered including via RF and optic fibre cables, across RF or optical free-space links, and synchronisation using the direct radar emissions or clutter returns. GPS time transfer is identified as uniquely suited to portable bistatic applications where a LOS breakthrough path may be unavailable. GPS time transfer is also cheap and can be autonomous which simplifies the passive node design. GPS time transfer is reviewed extensively focussing on the low-cost one-way method for which low-cost consumer timing receivers are readily available. The focus is on the published results for the receiver-to-receiver performance across short baselines of tens of kilometres. Then, follows a review of GPSDO technology which includes various GPSDO architectures and feedback strategies. Some reported time synchronisation mechanisms and quick-locking PLL techniques are studied. The expected performance of GPSDOs using various classes of STALOs are also considered. It is noted that GPSDO manufacturer specifications may often be inaccurate and not well suited to bistatic radar applications. Finally, some published accounts of GPS synchronised bistatic radar are reviewed. To date, there exist only a few examples. Nonetheless, it is noteworthy that the first reported fully functional GPS synchronised network Doppler radar was demonstrated as early as 1993.

2.1   Advantages and Challenges of Bistatic Radar

The field of bistatic radar and more specifically multistatic and networked radar has received much interest in the recent years. Bistatic radar has been around for nearly 80 years. However, relatively recent technological advances in data transfer rates, computing and synchronisation are now making more complex bistatic systems practically possible. [1, 2] This section examines the advantages of bistatic, multistatic and networked radar over monostatic systems. However, it also considers the challenges and increased complexity of such bistatic systems focussing on the difficulty of space-time synchronisation.

The literature describes many variants of the bistatic radar. In contrast to a monostatic radar, the transmitter and receiver of a bistatic radar are geographically separated. This bistatic concept can be further extended to a multistatic system where there are multiple separate transmitters and receivers. [3] The transmitters and receivers may also be connected via a data network to form a networked (or netted) radar. The individual transmitters and receivers are then referred to as nodes. [4] These radar nodes can also function as individual systems where each node performs some initial data processing before it is sent to a central processor [36].

Much of the advantage of a multistatic radar rests on the various possible layers of cooperation among the radar nodes. For example, the nodes could exchange already processed range, angle and velocity data, or the raw phase data is coherently processed at a central processor. The multiple view angles provide better spatial diversity and increase the amount of extracted target information. The overall sensitivity is increased, and there is a higher probability of viewing a stealthy target from a high RCS angle. Simultaneously, the probability of obscuration is reduced. Target triangulation is also possible. Passive receive-only nodes can be entirely covert and harder to detect and jam. Also, increased redundancy is thought to make it less vulnerable to failure or attack. [3, 4] Multistatic systems can operate at higher unambiguous pulse repetition frequency (PRF) rates than monostatic radar [16]. Passive receive-only nodes also have the potential to be cheap and simple [2]. Moreover, it was estimated that the cost of one passive node of a demonstrated experimental networked weather radar is less than 3% of the cost of one transmitting node. The designers reported that the additional costs associated with the multistatic system such as synchronisation, communication, data transfer and data integration are significantly less than the cost of a single transmitting node. [17] More recently, intelligent cognitive radars were described where highly cooperative systems adapt in response to the environment [5].

A network radar can be coherent and take advantage of transmitted phase data, or non-coherent where only the signal envelope is used [3]. A non-coherent network radar may improve the monostatic radar signal-to-noise ratio (SNR) by up to a factor of N, where N is the number of nodes. An entirely coherent netted radar with perfect space-time synchronisation can improve the SNR by up to a factor of N^2. However, here it is assumed that all nodes can transmit and receive to and from any other node. Also, the node pulses are transmitted such that they interfere constructively at the target. [4]

It is clear that a multistatic system has many advantages. However, these advantages are dependent on the system geometry, the modes of operation, the level of node cooperation, and on the space-time coherence [4]. However, such radar geometries are significantly more complicated, and high bandwidth data links are required. Also, the space-time synchronisation required for node cooperation and data fusion adds significant complexity in comparison to a monostatic system. Especially, for phase-coherent systems. [1, 2, 6] Moreover, the transmitter must communicate the carrier frequency, bandwidth, PRF, waveform and modulation to all receiving nodes. The positions of all nodes must be known, and the antenna beam patterns must be coordinated to allow common areas to be illuminated and observed simultaneously. [7] The exact pulse transmission time must be known network-wide to perform range measurements. Also, the carrier frequencies and phases of the nodes must be synchronised for Doppler estimation and coherent integration, respectively. [7–9] Finally, the phase noise requirements of the reference oscillator in bistatic radar is much more stringent than in monostatic systems. [10]

Thus, bistatic and multistatic systems may offer significant practical, tactical, and performance advantages over that of monostatic radar. However, the implementation of such systems is significantly more complex. The following sections consider the time, frequency and phase synchronisation requirements of bistatic systems, as well as, various synchronisation methods.

2.2   Bistatic Synchronisation Requirements

This section reviews the literature dealing with the time, frequency and phase synchronisation of bistatic radar. The synchronisation requirement is often processing dependent. However, this text is primarily concerned with carrier synchronisation before any processing. The spatial synchronisation requirements which include antenna beam coordination, motion compensation, and node positioning, are not within the scope of this text.

Node clocks can be directly, or continuously, synchronised. Alternatively, it can be done indirectly where the independent STALOs are synchronised intermittently. These STALOs keep the synchronisation indirectly during the hold-over periods.[3, 7] Note that the direct synchronisation requirement is equivalent to the relative Tx-Rx synchronisation requirement. Indirectly synchronised matched and independent STALOs must perform a factor of two better than the specified direct (or relative) synchronisation. The following section will focus on indirect synchronisation. Refer to Appendix A where various oscillator stability measures are reviewed.

2.2.1   Time

Relative node-to-node time offsets result in bistatic range errors [3, 7–9, 20] such that

$$\Delta R_B = c\Delta\tau_B, \qquad\qquad [\text{m}] \quad (2.1)$$

where ΔR_B is the bistatic range error due to the Tx-Rx time error $\Delta\tau_B$, and c is the speed of light. The typical time accuracy requirement is often a fraction of the transmitted compressed pulse width [3]. If one assumes the time accuracy requirement is one-tenth of the compressed pulse, then from [20] the time accuracy requirement can be expressed as

$$\Delta\tau_B = \frac{\tau_{pulse}}{10} = \frac{1}{10B}, \qquad\qquad [\text{s}] \quad (2.2)$$

where $\Delta\tau_B$ is the required node-to-node bistatic time accuracy, τ_{pulse} is the compressed pulse width, and B is the pulse bandwidth. Now from [3, 20] the required frequency stability is

$$\sigma_y(\tau_u) = \frac{\Delta \tau_B}{2\tau_u}, \qquad [\ ] \quad (2.3)$$

where $\sigma_y(\tau_u)$ is the required individual oscillator frequency stability[1] averaged over a time period τ_u, where τ_u is the rate at which the two clocks are re-synchronised. For the direct case, the above equation is multiplied by a factor of two where $\sigma_y(\tau_u)$ then represents the required relative node-to-node frequency stability. [3, 8]

Bing-Yang and Yan-Heng [37] also proposed an interesting use of the concept of geometric dilution of precision (GDOP) to measure the relationship between time accuracy and range accuracy in forward scattering systems. This may also be useful in other applications.

Equation 2.3 assumes perfect time and frequency synchronisation with no oscillator ageing (linear frequency drift), and that the time drift is solely attributable to frequency instability. However, the author is of the opinion that this is not always a realistic assumption for indirectly synchronised oscillators and short hold-over periods. In practice, relative frequency offset and ageing may dominate timing error.

2.2.2 Frequency

The target Doppler error due to imperfect frequency synchronisation is given by the integral of the Tx-Rx frequency offset over the coherent integration time. [9] A constant node-to-node carrier frequency offset of Δf results in a target Doppler error of Δf. Fluctuations in the carrier frequency offset cause Doppler spreading over the integration period. [9] The above implies that the frequency offset and drift requirements are processing and application specific.

2.2.3 Long-Term Phase Stability (Phase Drift)

The bistatic phase stability specification is identical to a monostatic system. Moreover, the required phase stability depends on both the duration and technique of coherent processing. The required RF carrier stability across the integration period can range from less than a degree to tens of degrees. [3]

From [3, 8] if the allowable phase error is $\Delta \phi$, then the required matched local

---

[1]Refer to Appendix A for more on oscillator frequency stability measures.

oscillator stability (indirect method) is

$$\sigma_y(\tau_{int}) = \frac{\Delta\phi}{2\pi v_c \tau_{int}}, \qquad [\,] \quad (2.4)$$

where $\sigma_y(\tau_{int})$ is the required individual oscillator frequency stability averaged over the integration period τ_{int}, and v_c is the carrier frequency. Again, (2.4) assumes perfect time and frequency synchronisation with no linear frequency drift, and that the phase drift is solely due to frequency instability. Again, the author is of the opinion that this is not a realistic assumption in practice for indirectly synchronised oscillators and typical integration periods of a few seconds or less.

Also of note, Hurley et al. [36] investigated the effects of synchronisation on the SNR of a fully coherent networked radar. They studied how the transmitted pulses interact at the target. The pulses from the various nodes are transmitted such that they interfere constructively at the target site. The SNR decreased sinusoidally as pulse or phase synchronism is lost. However, this text focusses on simple bistatic pairs with no interference at the target.

2.2.4 Short-Term Phase Stability (Phase Noise)

The SCV of a pulsed-Doppler radar is a measure of its ability to detect small slow-moving targets in the presence of large clutter returns. The SCV is highly dependent on the close-in phase noise of the reference oscillator. [32] Noise far away from the carrier typically consists of white PM noise. Close-in phase noise refers to the noise close to the carrier where the dominating noise sources (for example flicker noise) increase with decreasing frequency.

In a monostatic radar, the up- and down-conversion are done by the same oscillator. Thus, the oscillator is not only deterministically synchronised (x_o, y_o, D), but the random phase fluctuations[1], $\epsilon_x(t)$, or phase noises, are correlated to a significant degree as well. This correlation causes a first difference operation on phase noise resulting in a significant phase noise cancellation. [10] However, in practical radar applications, there is a time delay between the transmitted and received pulses. Consequently, the transmitted phase fluctuations decorrelate from that of the receiver local oscillator (LO) during this time delay. This decorrelation results in a reduced cancellation of close-in phase noise during down-conversion. This increase

---

[1] Refer to Appendix A for more on deterministic and non-deterministic oscillator effects.

in close-in phase noise deteriorates the SCV. Moreover, it reduces the probability to detect small moving targets in the presence of clutter or a large target. [32]

Goldman [32] modelled this phase noise cancellation effect using a delay line and derived the 'delay function' from narrowband frequency modulation (FM) theory as

$$K^2(\omega_m) = 2[1 - \cos(\omega_m \tau_{delay})]. \tag{2.5}$$

Here, phase noise cancellation, $K^2(\omega_m)$, is a function of frequency offset from the carrier, $\omega_m = 2\pi f_m$, due to a time delay, τ_{delay}.

Auterman [10] also quantified this phase noise cancellation but went further to compare the resultant phase noise of monostatic and bistatic Tx-Rx pairs. Moreover, he modelled the effect on phase noise using a simple up- and down-conversion stage. From [10] the phase noise cancellation effect at the monostatic node can be expressed as

$$S_{n_M} = 2M^2[2\sin(\pi f_m \tau_{delay})]^2 S_\phi(f_m), \tag{2.6}$$

where S_{n_M} is the resultant phase noise at the receiver, M is the ratio of the STALO to the carrier frequency, $S_\phi(f)$ is the single sideband STALO phase noise.

Auterman's cancellation factor $[2\sin(\pi f_m \tau_{delay})]^2$ in (2.6) and Goldman's 'delay function' in (2.5) are derived using very different methods. However, the two are identical when recognising the trigonometric identity $2\sin^2\theta = 1 - \cos 2\theta$.

This monostatic phase noise cancellation factor is plotted in Figure 2.1 for delays of 200 us, 20 us, and a 100 ns, representing monostatic target ranges of 30 km, 3 km and 15 m, respectively. From the plot, phase noise cancels at a rate of 20 dBc/decade for decreasing offset frequencies starting at $f_m = 1/(2\tau_{delay})$. For a target at a monostatic range of 3 km, the close-in phase noise is suppressed by nearly 80 dBs at 1 Hz from the carrier. Also, the phase noise is increases by 6 dB at frequencies greater than $f_m = 1/(2\tau_{delay})$.

Auterman [10] further derived the resultant bistatic phase noise by assuming that two matched independent STALOs are used for up- and down-conversion. The bistatic phase noise is then given by

$$S_{n_B} = 4M^2 S_\phi(f_m). \tag{2.7}$$

Hence, there is no cancellation of phase noise or spurious signals in bistatic radar.

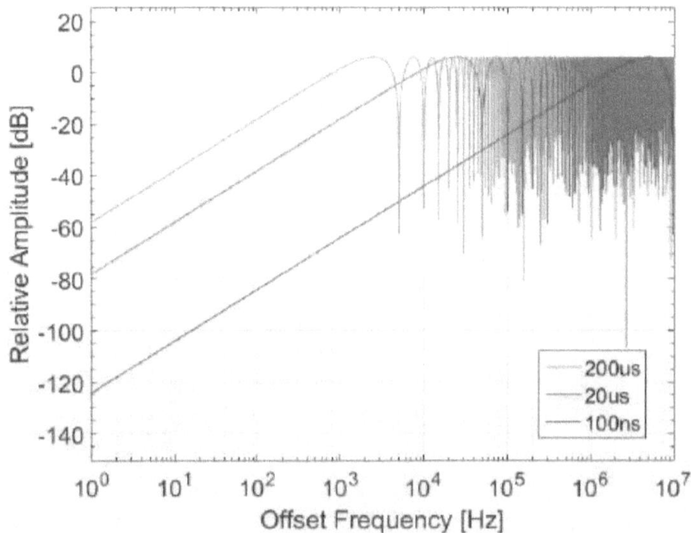

Figure 2.1: Monostatic Phase Noise Cancellation for Various Time Delays. The monostatic time delays of 100 ns, 20 µs, and 200 µs represent target ranges of 15 m, 3 km and 30 km, respectively.

To the contrary, the phase noise of the two matched STALOs adds. Thus, bistatic systems have a much more stringent phase noise requirement than monostatic systems. [10, 38] Especially, if a similar bistatic SCV is required as for monostatic radar. The effects of phase noise on BSAR systems are further developed in [10, 38].

2.3 Bistatic Synchronisation Methods

The focus of this text is primarily on low-cost GPS time transfer to synchronise bistatic radar. However, this section briefly reviews the alternative methods of time transfer, including via a copper or optic fibre cable, via a free-space RF or optical link, or using the transmitted radar emissions. This review is not exhaustive. The purpose is to highlight the advantages, challenges, and the expected performance of each method. Examples are also provided where each method was used to synchronise bistatic or multistatic radar systems. GPS time transfer is reviewed separately in Section 2.4. Refer to Willis [3] and Weiss [20] for excellent reviews of bistatic time and frequency synchronisation.

2.3.1 Cable

Optic fibre and copper cable can achieve time accuracies of 0.1 ns to 0.5 ns and 1 ns to 10 ns, respectively over shorter distances (circa 2004) [20]. However, it requires expensive and permanent infrastructure which is very limiting to portable networked radar. Especially over long baselines. Copper cable is also subject to signal dispersion and temperature induced phase changes [17]. Long stretches of a buried cable may also be difficult to protect and may be vulnerable to enemy tampering. Kesheng [14] reported on a multistatic radar system where either optic fibre or microwave links synchronised the Rubidium STALOs in a master-slave fashion. This particular system also used GPS as a fall-back strategy. They reported time and phase synchronisation of ±2 ns and 3.6°, respectively. However, the carrier frequency and baseline distances were not given. Fibre is also widely used to synchronise large radio telescope arrays. The ALMA radio telescope in Chile is one such example where femtosecond synchronisation is achieved across baselines of up to 18 km. [39]. Moreover, fibre optics are expected to achieve femtosecond time stability across 1000 m and picosecond stability across several kilometres [40]. However, this type of clock distribution is highly specialised and expensive. CERN recently developed a novel Ethernet-based fibre synchronisation system called White Rabbit (WR). This system is open source, commercially available, and relatively low-cost. It offers subnanosecond timing, and the fibre infrastructure carries both the synchronisation and data. Thus, appearing very promising to networked radar. [41, 42]. An ultra long distance WR link of 500 km was recently demonstrated with picosecond timing and a frequency stability of $\sigma_y(\tau = 2 \times 10^5) = 2 \times 10^{-15}$ [43]. In another experiment, WR was tested over a distance of 950 km [44].

2.3.2 RF Link

Time synchronisation across an RF link is a common synchronisation method. This method requires no cable infrastructure. However, microwave links often require direct LOS and are susceptible to interference, multipath and propagation effects

[45]. RF links are also vulnerable to enemy jamming and detection, and their performance is dependent on the received SNR [11]. For time synchronisation across an unknown baseline, two sites must exchange clock information to cancel the unknown propagation delay [45]. Schmid et al. [46] recently demonstrated a novel one-way (or open-loop) time transfer technique over a distance of 85 m. However, this requires continuous measurement of the baseline. Today there exist many techniques to perform time, phase and frequency synchronisation over RF. [47–49] Specialist RF links can achieve time synchronisation of between 10 ps and 100 ps [40]. A recent publication reported on a sub-50 ns level implementation of WR across a microwave link [50]. Bistatic synchronisation across RF requiring Tx-Rx LOS are often airborne or spaceborne systems which are less susceptible to multipath and interference than ground-based systems. The TanDEM-X interferometer [11] is an example of a spaceborne X-band BSAR system where precise synchronisation is achieved using two-way synchronisation across RF. They achieved a phase error of less than 1° over baselines of up to 1 km. He et al.[51] also elaborately discuss phase synchronisation of distributed spaceborne frequency modulated continous wave (FMCW) synthetic aperture array (SAR) across RF. This scheme uses separate Tx-Rx antennas for phase synchronisation. Time transfer via two-way satellite communications can achieve a time synchronisation of between 1 ns and 10 ns and alleviates the need for direct LOS. However, such satellite communication requires expensive equipment. [20, 45]

2.3.3 Optical Link

Free-space optical links are showing promise for femtosecond level time synchronisation of high-performance optical oscillators for future distributed sensors. Moreover, a two-way synchronisation technique was demonstrated across up to 4 km with a 40 fs peak-to-peak wander over two days and $\mathrm{Mod}_{\sigma_y}(\tau > 10^3) \approx 2 \times 10^{-19}$. This technique proved rather resistant to strong turbulence and changes in weather. [40]

2.3.4 Radar Transmissions

Bistatic radar has been synchronised using the transmitted radar emissions since the early 1980's and likely even earlier. It has been demonstrated using the direct signal, the sidelobe breakthrough, or a common target or clutter reflections. However, these methods are subject to multipath, interference, propagation effects, and often require cooperation between the bistatic pairs. [3] RABIES [9, 15], an experimental bistatic system of 1987, achieved a timing accuracy of $\pm 5\,\mu s$ using the sidelobe breakthrough over baselines of tens of kilometres. Another experimental bistatic system [16], also of 1987, used the tropospheric reflected breakthrough to synchronise beyond the LOS over a baseline of 200 km. A 2.8 GHz Doppler weather radar network (circa 1993), demonstrated sidelobe synchronisation across baselines of 10 km to 20 km. This system achieved time and frequency synchronisation of sub-1 μs and better than 10^{-9}, respectively. [17] Retzer [7] demonstrated a system where a separate receiver channel recorded the direct signal. This reference signal was then cross-correlated with the reflected pulse to correct the time and frequency offsets. Many non-cooperative passive radars use a similar approach today. However, most still require some form of external synchronisation. Saini, Zuo, and Cherniakov [21] described an space-surface bistatic SAR (SS-BSAR) system where GLONASS was used as an illuminator of opportunity. Here the passive node recorded and decoded the direct signal whereafter it is used for both the SAR imaging as well as time and frequency synchronisation. See [52] for an extensive analysis of non-cooperative bistatic synchronisation. A novel method was also recently published where a chaotic bistatic radar is synchronised using the transmitted waveforms. [53, 54]

2.4 GPS Time Transfer

The previous sections reviewed alternative methods of time transfer, including via copper and fibre cables, via RF links, and using the radar transmission. However, GNSS offers a mode of time transfer which is perhaps uniquely suited to portable network radar synchronisation. Moreover, GNSS is cheap, portable, completely covert, can be completely autonomous, and it does not depend on LOS or the transmitted radar signals. This section reviews GNSS time transfer (specifically GPS) as a method to synchronise networked radar systems.

GNSS time transfer (specifically GPS) has become increasingly popular since it became available in 1992 [18]. It is always available free of charge nearly any-where on earth and has become the primary method of time transfer [19]. The only infrastructure required is a GNSS timing receiver. It does not require LOS between the radar nodes, it can be autonomous, and it simultaneously provides an accurate means to determine the spatial coordinates of each node. GNSS receivers are passive and thus not at risk of enemy detection. However, they are subject to enemy jamming and spoofing. Nonetheless, GNSS offers a solution to the challenge associated with the time and synchronisation of bistatic radar [1]. Today there are various GNSS systems such as the Russian GLONASS, the European Galileo, and others. Multiple systems offer redundancy, and future systems are expected to offer improved performance [21]. This document will focus on the United States owned Navstar GPS. Refer to [55] for a more in-depth review of the GPS system. Consult the U.S. Space-Based Positioning, Navigation, and Timing (PNT) Policy of 2004 [56] for more information.

The GPS satellites transmit navigational signals on two carriers. The L1 and L2 carriers are at 1575.42 MHz and 1227.60 MHz, respectively. Civilian use is limited to the C/A PRN codes of the L1 carrier. The military encrypted precision (P(Y)) PRN codes are transmitted on both the L1 and L2 carriers. The military code has a bit rate that is ten times higher, and the synchronisation is a factor $\sqrt{10}$ more precise [57]. The future GPSIII is expected to have an L5 carrier with improved performance [21]. However, the civilian GPS C/A code was initially superimposed with a random dithering signal to degrade its accuracy. Nevertheless, averaging over long periods of time improved the short-term timing performance significantly. Atomic oscillators allow averaging times of up to 24 hours or more. However, in May 2000 the S/A directive was switched off affording civilians the full accuracy of the L1 C/A signal [26]. Further, in September 2007, it was announced that the new GPSIII system would be without the S/A function [58]. Removal of this intentional dithering improved the RMS time jitter produced by civilian receivers by a factor of five. Cheaper oven-stabilised quartz oscillators that are orders of magnitudes less stable can now deliver levels of synchronisation previously only possible with expensive Rb atomic sources. [25] Hence, atomic level frequency stability became available to the low-cost market when S/A was switched off.

GPS time transfer offers low-cost, autonomous, and accurate time and frequency synchronisation which appears to be suited to low-cost coherent networked radars. However, before it could be applied to bistatic radar, it is essential to understand its practical and performance limitations. The following sections review the different modes of GPS time transfer. It is explored how specialist GPS timing receivers differ from the more common navigation receivers, and also identify the various sources of GPS timing errors. Finally, the timing performance of the three GPS time transfer techniques is compared. The receiver-to-receiver one-way performance results for the low-cost Motorola M12+ timing receiver across a 21.5 km baseline are included. Moreover, for short baselines, the receivers track the same constellation and are equally affected by the atmosphere. Thus, suggesting that one-way receiver-to-receiver timing performance is significantly better over short baselines than what is quoted across large baselines to UTC. This increased relative performance across short baselines is of particular interest to low-cost, short baseline, bistatic systems.

2.4.1 Time Transfer Methods

There are three GPS time transfer methods, namely the one-way, or direct broadcast method, common-view method, and the carrier phase method. The three are techniques summarised below from [18, 19, 23].

The one-way method is the most widely used method and is completely autonomous. Civilian receivers use the L1 C/A codes broadcasted directly from the satellites to compute its position. A single satellite or multiple satellites can be used. One-way is the simplest method of time transfer. However, it may not have adequate timing performance for high-resolution bistatic applications.

The common-view method requires two-way communication between two receivers and some post-processing. Two civilian receivers use the L1 C/A codes from a satellite in common-view and compare the GPS time to a frequency standard at each of the two sites. Then, the measured time errors are exchanged and subtracted. The errors that are common to both sites, such as the atmospheric errors and satellite clock bias, are cancelled during this difference computation. A single satellite or multiple satellites can be used. The common-view method is most useful for baselines shorter than a few thousand kilometres. For baselines over 10,000 km, the timing performance approach that of the one-way method. Laboratories often use this technique to steer their local clocks to UTC time. Although this technique is more complex and requires node cooperation, it may not necessarily be more expensive since node-to-node communication is often already available in networked radars.

The carrier phase method is much more accurate than either the one-way or common-view methods. It utilises the phase of both the L1 and L2 carriers instead of the codes along with precise models of the satellite orbits, troposphere and ionosphere. This technique requires extensive post-processing and can itself operate in one-way or common-view modes. This technique is not yet in mainstream use today due to the associated cost and complexity. Thus, it may not be a practical choice in most bistatic application. However, future carrier phase GPS time transfer techniques must be researched with a high priority because it could potentially offer low-cost sub-nanosecond timing.

2.4.2 Low-Cost Timing Receivers

Timing receivers are different from navigation receivers in a number of ways. A timing receiver is used stationary with its position precisely surveyed. The receiver then assumes this fixed position and solves for time only. Thus, only a single satellite is required to acquire a time fix, and solutions from the multiple tracked satellites are averaged to improve the timing performance. [19] Since the receiver is stationary, the received signals do not have Doppler offsets due to relative receiver velocity. Thus, GPS timing antennas and receivers typically have much narrower out of band hardware filtering in comparison to conventional navigational antennas and receivers. This improves interference immunity. Such receivers may also use adaptive tracking loops with narrower bandwidths to improve interference immunity further. [59] A GPS timing antenna is usually active with a built-in low-noise amplifier (LNA) [19]. Antennas should be mounted high to minimise multipath with an unobstructed view of the skies. The default antenna mask angle is set between 10° and 20° to avoid satellites that are close to the horizon. Signals from low elevation satellites are noisier due to the longer propagation through the atmosphere. [59]

Low-cost OEM timing receivers are available in many forms ranging from credit card sized printed circuit board (PCB) modules down to much smaller chip scale modules. The Motorola M12+ timing GPS receiver [59] is an example of a credit card sized OEM module. It is a low-cost multi-channel consumer timing receiver and is often paired with the Motorola Timing 2000 active antenna. The M12+ timing receiver includes survey and position hold modes as well as the time-receiver autonomous integrity monitoring (T-RAIM) algorithm. In survey mode, it estimates a more accurate position by averaging 10,000 position estimates in an attempt remove the zero mean aberrations. In position hold mode, it assumes the averaged position estimate as a known and fixed position. This mode uses all satellites to solve for time, and it calculates the time average for improved accuracy. The T-RAIM algorithm uses this redundancy to detect and remove satellites that are producing incorrect time values. [59]

Like most other low-cost timing receivers, the M12+ delivers the time in the

Table 2.1: GPS Time Transfer Error Sources (nominal RMS values). [23]

Error Source	Direct Single-Frequency GPS UTC(USNO)_Lab_Y	Common-View GPS Lab_X_Lab_Y	Carrier Phase GPS Parameter Solution Lab_X_Lab_Y	TWSTFT Lab_X_Lab_Y
Multipath Bias (24-hour average)	1–3 ns	1–4 ns	1–4 ns	< 1 ns
Multipath Precision (13-minute observation)	1–10 ns	1–14 ns	< 20 ps	< 100 ps
Model Ionosphere	3–6 ns	3–6 ns*	0 ns	30–300 ps**
Troposphere at Zenith	300 ps (no fit)	500 ps (no fit)*	10 ps after fit	0
Broadcast Orbits	< 3 ns/observation	< 3 ns/observation*	—	< 100 ps*
Broadcast Clocks	< 3 ns/observation	0	—	—
Hardware Variations	2 ns per year	3 ns per year	3 ns per year	< 1 ns per year
Receiver Noise (averaged over 5 minutes)	< 1 ns	< 1 ns	< 20 ps	< 20 ps
Earth Tide (at any instant)	300 ps**	300 ps**	< 10 ps*	0 ps

* Baseline dependent, better for short baselines

** Assuming effect is not modeled. Baseline dependent for common-view and TWSTFT

form of a 1 Hz, or PPS, square wave signal. The PPS signal has exceptional long-term accuracy and stability that is traceable to UTC. However, the receiver's cheap temperature-compensated 16,367 MHz onboard oscillator is not a multiple of 1 Hz. Thus, the receiver cannot reproduce a 1 Hz signal exactly and superimposes a time quantisation error onto the PPS signal. This zero-mean error is known as the sawtooth error and has a peak-to-peak value of 30 ns which corresponds to half the clock period. [60] The receiver provides the user with an estimate of its clock offset (sawtooth error) before the arrival of each PPS edge for removal in post-processing [59]. The raw and sawtooth correct PPS time error of an M12+ is compared in Figure 2.2 [60]. Sawtooth compensation reduces the high-frequency noise to a few nanoseconds. The remaining jitter is largely due to ionospheric and receiver noise [23].

In many practical applications, the GPS PPS output is not useful by itself. Applications often require better short-term stability and a higher reference frequency of 5 MHz or 10 MHz. In such cases, it is customary to steer a STALO using the GPS PPS signal. Thus, the STALO delivers the required short-term stability, and GPS compensates for the longer-term STALO instabilities such as frequency offsets and drifts. Such a system is referred to as a GPSDO. [19, 23] Refer to Section 2.5.

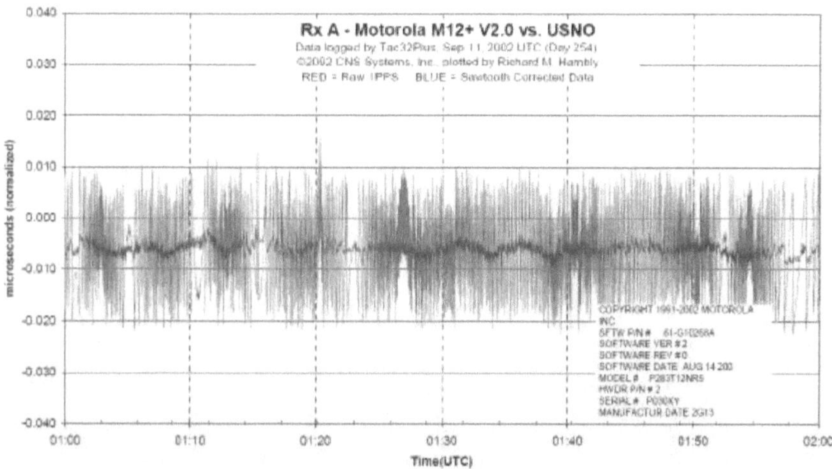

Figure 2.2: The M12+ PPS Output: RAW versus Sawtooth Corrected. Sawtooth correction improves the jitter from roughly $\pm 15\,$ns to only a few nanoseconds. [60]

2.4.3 Timing Error Sources

Numerous error sources affect the precision, accuracy and stability of a GPS receiver. These sources are summarised in Table 2.1 for the different time transfer methods.

The atmospheric noise, clock and orbit biases, and receiver noise are mostly out of the control of the user. However, it is possible to cancel the atmospheric noise and clock biases partially by using the common-view method across shorter baselines ($\leq 2500\,$km). On the other hand, the user has partial control over the timing errors due to antenna, receiver and cable biases, multipath, survey errors and temperature variations. Parker and Matsakis [23] noted that the calibration of a system could wander by several nanoseconds over months and years. Also, multipath can introduce errors of several nanoseconds. Inaccurate surveyed antenna coordinates cause a time offset. [31] Often, the GPS receiver's built-in auto survey function is the best available alternative to a professionally surveyed antenna position. However, averaging 10,000 position estimates takes nearly 3 hours to complete. Lombardi, Novick, and Zhang [31] evaluated the auto survey function of four commercial GPS receivers. The auto-surveyed longitudes and latitudes were accurate, but the altitude error ranged from 1.7 m to 23.2 m across the receivers for a single test. An altitude error results in a time error at a rate of 3 ns per meter [31]. Thus, the resultant time errors would have ranged from 5.1 ns to 69.6 ns. Ambient temperature changes affect signal propagation through the receiver system causing time errors of 2 ns to 3 ns [61]. Lisowiec et al. [61] successfully demonstrated a low-cost temperature stabilisation system for the Motorola Timing 2000 antenna. In summary, the user can optimise the timing performance in several ways. The receiver system must often be calibrated, and the antenna must be accurately surveyed and carefully mounted. Temperature stabilisation may also improve the time error.

2.4.4 Timing Performance

Today, consumer level single carrier multi-channel GPS timing receivers are widespread and commercially available. These low-cost GPS timing receivers offer remarkable timing performance extensively documented in the literature. The expected performance figures for the various time transfer methods are summarised below from [18, 19].

Table 2.2 compares these GPS time transfer methods for 24-hour averaging periods. The one-way method delivers excellent long-term time accuracy and frequency stability of less than 20 ns and below 2×10^{-13}, respectively. However, it has a significant high-frequency noise component with a frequency uncertainty of a few parts in 10^{-9} averaged over one second, and a peak-to-peak time error of about 50 ns for 10-minute data point averages. The single-channel common-view technique improves this timing performance by up to a factor of two over the one-way method. This uncertainty could be further halved by using multiple channels. Hence, providing a factor of four improvement when compared to the one-way method. Carrier-phase time transfer, on the other hand, is approximately twenty times more accurate and stable than one-way time transfer [19].

Averaging periods of a day or more requires an ultra-stable atomic source to filter the signal. However, the stability of quartz references limits the possible averaging periods to a few thousand seconds or less. Nonetheless, removal of the S/A dithering from the GPS signal reduced the RMS time jitter by a factor of five, and quartz-based GPSDOs are now able to deliver stability levels previously only possible using more expensive Rubidium STALOs.

Refer to the frequency stability plot for one-way time transfer in Figure 2.3. Stability plots for common-view and carrier phase time transfer can be found in

Table 2.2: Performance Comparison of GPS Time Transfer Methods [19]

Method	Time Uncertainty 24h, 2σ	Frequency Uncertainty 24h, 2σ
One-Way (multi)	$< 20\,\text{ns}$	$< 2 \times 10^{-13}$
Common-View (single)	$\approx 20\,\text{ns}$	$\approx 1 \times 10^{-13}$
Common-View (multi)	$< 5\,\text{ns}$	$< 5 \times 10^{-14}$
Carrier-Phase	$< 0.5\,\text{ns}$	$< 5 \times 10^{-15}$

[19]. Note that the frequency stability improves linearly with increasing averaging periods for all methods.

The one-way results quoted above are absolute deviations from UTC over large baselines. However, this text is concerned with receiver-to-receiver synchronisation over very short baselines of a few kilometres or less. Thus, the one-way multi-channel results published by Hambly and Clark [60] is of particular interest. First, they calibrated the receivers by simultaneously comparing the receiver-to-receiver tracking of four colocated Motorola M12+ timing receivers. Then, they measured the receiver-to-receiver tracking across a baseline of 21.5 km each locally compared to a hydrogen maser.

The zero-baseline relative time errors are plotted in Figure 2.4. The receivers tracked each other closely, but roughly 5 ns offset two of the receivers. The authors confirmed that a manufacturing tolerance in the receiver IF filter caused this 5 ns offset which highlights the importance of receiver calibration. The plot of several days shows that the receivers are prone to 'glitches' ranging from tens to a hundred nanoseconds. There are also slower excursions lasting up to 6 hours of up to 50 ns which will be more difficult to filter using shorter averaging periods.

Two M12+ receivers are compared across a baseline of 21.5 km in Figure 2.5. The plot shows 10-minute averages over several days. The peak-to-peak time difference was on the order of ± 15 ns. However, smoothing reduces this time error to less than ± 5 ns. Of particular interest is the *aurora borealis* event which caused the large 40 ns transient towards the end of the third day. Here, both receivers were affected equally which left the relative time difference unaffected. Moreover, for short baselines, the receivers are tracking the same satellites and are equally affected by atmospheric effects. Thus, suggesting that one-way receiver-to-receiver timing performance is significantly better over short baselines where individual receivers track the same

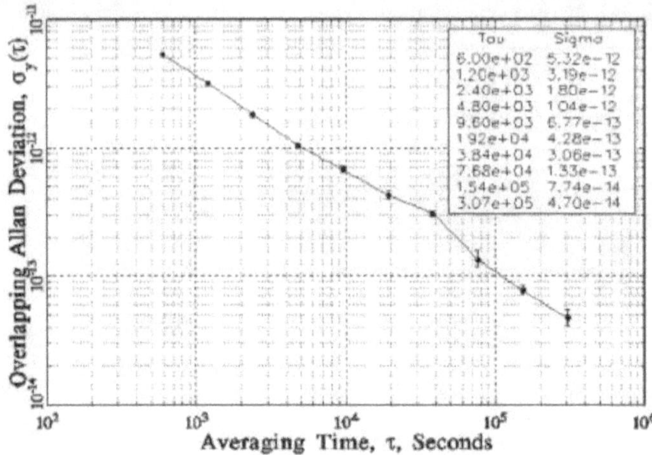

Figure 2.3: Frequency Stability of GPS One-way (Multi-Channel).[19] Receiver versus UTC(NIST).

constellation than what is quoted across large baselines to UTC.

2.5 GPS Disciplined Oscillators

GPSDOs offer remarkable time and frequency synchronisation at low cost and seem particularly suited to bistatic, multistatic, and networked radar synchronisation. GPSDOs exploit the long-term stability of GPS, or other GNSS, to correct the time, phase and frequency drift effects caused by ambient changes and STALO imperfections. The result is an oscillator possessing both the short-term STALO performance and the long-term GPS stability. These self-calibrating systems are relatively immune to ambient changes with long-term stabilities comparable to that of Cesium. However, they have the added benefits of being cheaper and traceable to UTC. GPSDOs use either quartz OCXOs or more stable and more expensive Rubidium STALOs depending on the budget and application.[22, 24] Refer to Lombardi [22] for an excellent introduction to GPSDOs.

GPS is the most commonly used GNSS for GPSDOs. However, the Russian GLONASS, the European Galileo, and the Chinese BeiDou [62] systems have also

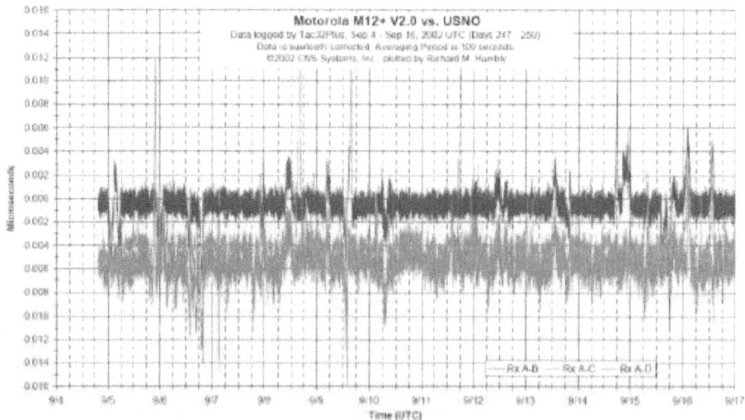

Figure 2.4: Comparison of Four Motorola M12+ Receivers (Zero-Baseline). [60]

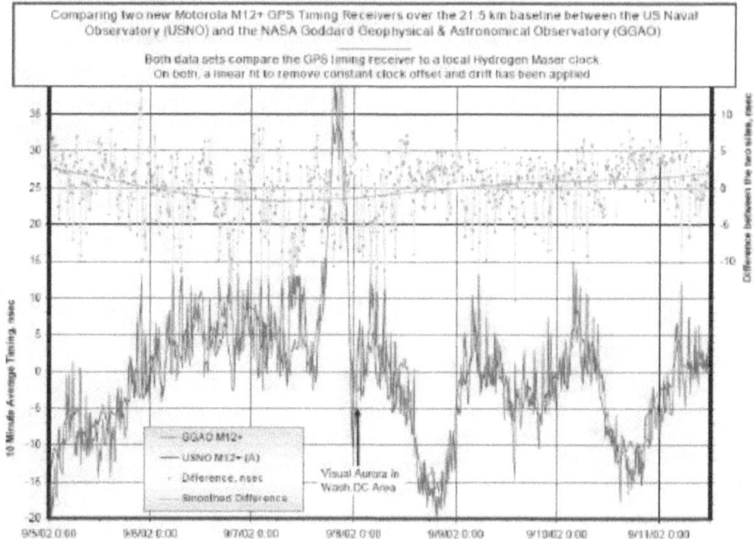

Figure 2.5: Comparison of two Motorola M12+ Receivers (21.5 km Baseline). [60]

since become available offering welcome redundancy. [21] However, this text limits itself to the US DoD owned GPS.

External oscillator steering for improved hold-over existed since the sixties, and possibly earlier, when Allan et al. [63] described STALO drift compensation based on electro-mechanical ball disc integrators. Later, as early as 1990, and before GPS became available to the public, a sophisticated common-view GPS disciplined system was described [24]. Today, GPS is the defacto standard for oscillator calibration in many laboratories [19]. Throughout the years, various GPSDO designs appeared in the literature [24, 25, 30, 62, 64–73]. The amateur community also produced, and continue to produce, well-documented GPSDO designs [74]. Even so, specific literature remains relatively sparse, and commercial GPSDO designs are often closed and proprietary [22]. Further, commercial GPSDOs may often behave unexpectedly with their datasheet specified performances only an estimate of their actual performance [31].

GPSDOs are often used in ideal laboratory environments with precisely surveyed antennas [31]. In such applications, loss of GPS lock is uncommon and excessive GPSDO lock-in times may not be a hindrance. However, portable multistatic systems may require regular geometry changes making the deliberate or accidental loss of GPS lock much more likely. Long GPS lock-in times of tens of minutes to several hours are very limiting in such applications. However, very few examples of quick-locking GPSDOs exist in the literature [71, 73]. Additionally, bistatic radar applications require exact time synchronisation. However, many commercial GPSDOs merely offer the jittery GPS PPS signal for time synchronisation whereas in other cases, the GPSDO PPS output is not in phase with its STALO output [31]. Only a few examples of GPSDO time synchronisation exist in the literature [30, 70, 72].

This text focuses on low-cost GPSDOs based on commercially and widely available one-way single-carrier GPS time transfer. The following sections briefly review the general architecture of a GPSDO. After that, the various feedback control strategies and the published quick-lock techniques are discussed. Then, the published time synchronisation strategies are reviewed. Finally, the relevant aspects of GPSDO performance are analysed.

2.5.1 Architectures

A GPSDO measures the time or frequency difference between a STALO and a GPS reference and then steers its output frequency to minimise the difference and optimise stability. It is a special kind of frequency synthesiser generating a high-frequency (1 MHz to 100 MHz) output from the very low frequency (1 Hz to 100 Hz) GPS time reference.

Various GPSDO architectures have been described over the years. The phase detection strategies range from the more traditional digital circuits such as XOR gates and flip-flops [71, 74] to very elaborate low-noise analogue quadrature detectors [73]. Phase detection can also be accomplished using DSP [75, 76] and time-to-digital converter (TDC) [71] techniques. The frequency compensation is often accomplished through direct STALO steering using a high-resolution digital-to-analogue converter (DAC). However, the STALO may also be free-running with the output steered digitally using a synthesiser [22, 72]. Wang [76] described a high-resolution frequency steering technique where the filtered DDS output is mixed with the original STALO. This mixing process increases the frequency resolution by a factor of 10^7 where the STALO frequency is 10 MHz. The STALOs used are usually either quartz OCXOs or Rubidium atomic standards depending on the application and budget [22]. The feedback control and filtering are usually done digitally within a microcontroller or field programmable array (FPGA). Analogue filters are subject to environmental and component drift effects [73]. Moreover, it is important to keep the short-term noise contribution of the phase detection and steering below that of the STALO. Green [73] provided an excellent account of the required low-noise design when using an ultra-low noise quartz Boitier a Vieillissement Ameliore (BVA)[1] STALO. GPSDOs have also been ruggedised against large temperature gradients and vibration. Vibration resistance may be important in shipborne or airborne radar applications and can be accomplished through active accelerometer-based compensation [72]. Thus, GPSDO architectures may take on many forms based on the budget, required performance, and specific application.

---

[1]Enclosure for improved ageing.

2.5.2 Feedback Control Strategies

The negative feedback control algorithms accomplishing the GPSDO phase- or frequency-lock leave much room for innovation and may be tailored to various applications.

A GPSDO may use various kinds of negative feedback control. Most are PLL-based [22], but they may also be Kalman-based [65]. PLLs and frequency synthesis by phase-lock are well-understood concepts [77–79], and a recent tutorial offered an insightful comparison between PLL and Kalman feedback techniques [80]. Kalman filters are computationally more intensive than PLLs but may offer certain advantages. For example, they can be arbitrarily frequency offset whereas a PLL cannot. [65] Offsetting the output frequency may be convenient to null the bistatic radar carrier bias caused by the finite synthesiser resolution at each node. Kalman filters may also provide superior hold-over performance by modelling the ageing and temperature dependence of the STALO during a GPS outage. [64, 67] Hold-over temperature compensation of Rubidium STALOs is generally more successful than that of quartz STALOs which are known to have non-linear temperature coefficients [67].

PPS pre-filters are often used to reduce the GPS noise before it enters the feedback algorithm [64, 71, 73]. These pre-filters exploit the zero-mean near Gaussian properties of the PPS time mark and have much higher bandwidths than the control loops to ensure the feedback is unaffected [73]. Both moving average [71] and least squares fit (LSF) based [64] pre-filters were proven to be effective.

Optimising the control loop bandwidth is vital. In general, the optimal bandwidth is considered to be at the frequency stability cross-over point of the STALO and GPS reference. The STALO stability is often constant, whereas the GPS stability is a function of antenna position, interference, multipath, and atmospheric conditions. Hence, the optimal bandwidth is dependent on the GPS reference. [68] However, very long feedback time constants limit the control loop's ability to react to ambient changes [64]. A mistuned control loop may produce sub-optimal results [75]. However, Rochat, Leuenberger, and Stehlin [68] demonstrated a novel adaptive filter which auto-tunes the PLL bandwidth based on the stability of the reference input.

The GPSDO phase lock-in time is dependent on the STALO stability. GPSDOs with more stable STALOs generally have longer optimal time constants which increase the time to phase-lock. Time constants for quartz and Rubidium GPSDOs typically range from hundreds to a few thousand seconds and tens of thousands of seconds up to a day or more, respectively. Hence, GPSDOs tend to have extended phase lock-in times which increase for decreasing feedback bandwidth. However, shortening the GPSDO lock-in times is of particular interest to portable networked radar applications requiring regular geometry changes. Nonetheless, only a few publications discuss quick-locking strategies. One strategy is to pre-tune the STALO closer to the correct frequency before enabling the PLL using a previously recorded DAC setting [71, 73]. Also, some applications may only require frequency-lock and not phase-lock. FLLs have shorter lock-in times [65]. The FLL lock-in time can be reduced further by first setting the PID controller to a proportional only mode until it achieves a target synchronisation level [71]. A quick-locking PLL was described where the STALO phase is first locked using the maximum possible bandwidth. Once phase-lock is achieved, it switches to the optimal bandwidth setting [73]. Unfortunately, none of these articles provided much information on the performance of their quick-lock methods. There is also little published on GPSDO phase-locked indication.

Thus, much of the innovation within GPSDOs lies in their feedback strategies. Sophisticated control algorithms may offer superior performance with improved filtering, adaptive bandwidths, and quicker lock-in times. Reduced lock-in times are crucial to the practical use of portable GPS-synchronised networked radar systems. Also, active hold-over compensation may significantly improve the free-running GPSDO performance. Hold-over algorithms use oscillator models and real-time temperature data to steer the STALO during a GPS outage. However, relatively little details are published on smart GPSDO feedback control. Further, the feedback control of most GPSDOs is implemented digitally within a microcontroller or FPGA making the research and development difficult and time-consuming. Hence, a rapid development environment for GPSDO compensation algorithms is desirable.

2.5.3 Time Synchronisation

The STALO output of a phase-locked GPSDO is phase synchronous to the averaged phase of the GPS PPS reference and thus also time synchronous. However, because of the high GPS PPS jitter, it remains problematic to identify the STALO zero-crossing which best corresponds to the GPS second. Very few commercial GPSDOs offer low-jitter GPS synchronous PPS timing outputs. Most low-cost GPSDOs only offer the jittery GPS PPS for time synchronisation. Those offering a more accurate PPS output is often not phase aligned to the STALO output. [31] McClelland, Staerman, and Zarjetski [72] described a system where a DDS is used to adjust the output PPS phase to match that of the input phase to an accuracy of 0.8 ns. However, very little detail is given. Baojian, Dehai, and Dazhi [30] described a novel technique where the output of an FLL is delay line adjusted until both the STALO and PPS outputs are in phase with the GPS PPS. They used an 8-bit delay line with a 0.25 ns resolution. Yulin, Jianyu, and Jintao [70] reported on a time synchronisation technique which improved the PPS time accuracy from ±100 ns to ±25 ns using a weighted averaging process. However, it is not entirely clear exactly how this mechanism works.

From Hambly and Clark [60] the relative time error between two M12+ timing receivers may be on the order of ±15 ns for 10 minute averages and a baseline of 21.5 km. Further smoothing reduces this error to ±5 ns (Section 2.4.4). Conceptually a GPSDO performs real-time low-pass phase filtering or smoothing, and it should be possible to achieve a relative GPSDO time error similar to that of the smoothed PPS errors. However, it remains problematic to realise time synchronisation with a similar accuracy than the relative GPSDO phase. Hence, it is of interest to investigate time synchronisation methods for low-cost GPSDOs further.

2.5.4 Performance

The mode of GPS time transfer and the quality of its STALO are the critical cost and performance drivers of a GPSDO. However, the feedback and oscillator steering algorithm can significantly impact its phase behaviour and short- to medium-term performance.

Refer to Section 2.4.4 for the limits of GPS time transfer. This text is limited to GPSDOs employing one-way single-carrier GPS time transfer. The literature reports extensively on the performance of such GPSDOs. Moreover, various quartz and Rb GPSDOs are compared to Cesium (Cs) standards and each other [22, 25, 31, 81].

Standalone Cs references have excellent short-term stability and phase noise. However, their tubes have limited lifespans making their cost of ownership much more than that of GPSDOs. Cs references also require intermittent time calibrations whereas GPSDOs do not.[22] However, GPSDO short-term stability is often deteriorated by its active frequency steering [22, 31, 81]. Refer to Table 2.3 for a performance comparison between Rb-based GPSDOs and standalone Rb and Cs standards. This table compares laboratory grade units. However, low-cost quartz-based GPSDOs and some based on miniature Rb clocks may cost less than a $1000. In the more recent years, chip scale atomic clocks (CSACs) have also become available offering atomic stability at reduced cost, weight, size, and power consumption.[72, 82] However, CSACs compromise on phase noise and short-term stability [82]. BVA SC-cut OCXOs offer the best available short-term stability and close-in phase noise performance but are significantly more expensive than other SC-cut OCXOs [73, 81].

All good GPSDOs should achieve Cs-like long-term performance. A high-grade laboratory Rb GPSDO is expected to achieve $\sigma_y(\tau = 10^5) \leq 10^{-13}$.[1] [22] However, a well-controlled commercial Cs standard can outperform the long-term stability of GPSDOs by a factor of 5 to 10 [83]. Weiss, Lisowiec, and Steplewski [25] postulated that a high-quality OCXO-based GPSDO is limited to a medium-term stability of $\sigma_y(\tau = 10^3) \approx 10^{-11}$, a long-term stability of $\sigma_y(\tau = 10^5) \approx 4 \times 10^{-13}$, and a peak-to-peak time error of below 40 ns. The results published in [22, 31, 81] appear to confirm these suggested limits.

Most GPSDOs are specified for use in calibration laboratories. Such laboratories are often interested in the absolute time and frequency offset to UTC, and the frequency stability averaged for up to a day [22]. However, bistatic pulsed-Doppler applications are more concerned with the instantaneous Tx-Rx time and frequency accuracy at the time of target observation. Bistatic radar is also affected by the change in phase and frequency during the target observation, pulse integration, and Doppler integration. No published results for the instantaneous GPSDO-to-GPSDO phase and frequency across short baselines of a few kilometres were found. However, there are zero-baseline results published.[31]

Moreover, the instantaneous GPSDO frequency offset may be orders of magnitude higher than that of the long-term averaged values. The instantaneous frequency offsets of OCXO-based GPSDOs ranged between $\pm 7 \times 10^{-11}$ and $\pm 2 \times 10^{-9}$ [81]. Further, Lombardi, Novick, and Zhang [22, 31] found that different commercial GPSDO models may have very different phase behaviours. Johnsen [84] came

---

[1]There are 86400 seconds in a day which is located at approximately 10^5 seconds on a logarithmic scale.

to a similar conclusion. In one extreme example, the peak-to-peak phase variation over 80 days of one Rb GPSDO was measured at 38 ns whereas another Rb GPSDO wandered by 588 ns [31]. Also, the manufacturer specified GPSDO performances of various models were found to be mere estimates of their actual performances [31]. Typically, manufacturers do not specify the instantaneous GPSDO phase and frequency performance over short- and medium-term time scales. Hence, the manufacturer datasheets are often not well-suited to bistatic radar design.

Thus, there is a need for well-understood GPSDOs of known and predictable performance and behaviour for research on bistatic and multistatic synchronisation. It is of interest to characterise the relative instantaneous phase and frequency of such GPSDOs over the short- and medium-term.

2.6   GPS-based Radar Synchronisation

There is little published on the GPS synchronisation of bistatic, multistatic, and networked radar. Even less is published on the achieved performance and implementation of such systems. Wang [76] noted the same for literature dealing with BSAR, multistatic SAR, and distributed SAR (DiSAR) where publications often focus on compensation within signal processing rather than the implementation of synchronisation. A few published examples of where real bistatic radar systems have been synchronised using GPS are discussed below.

Shortly after GPS became available, Wurman et al. [17] pioneered networked radar GPS-synchronisation in 1993. An existing high-powered 2.8 GHz WSR-88D weather radar supplied illumination for multiple low-cost passive receivers with low-gain antennas. This dual-Doppler networked radar prototype demonstrated a cost and geometrical advantage over that of multiple high-powered monostatic radars. Both the sidelobe breakthrough, as well as, GPS, were used for time and frequency synchronisation. It used one-way OCXO-based GPS-disciplining. These OCXOs aged by less than 1×10^{-9} per day during hold-over. A telephone network was used for data transmission and control. They achieved a frequency offset of 1×10^{-9} over tens of kilometres. This radar combined the direct breakthrough and GPS to synchronise time. GPS was used to start the passive node sampling about 10 µs before the transmitted signal. Thus, the direct signal was recorded and then used for time synchronisation. Time synchronisation of 1 µs was achieved with a drift of less

Table 2.3: Performance Comparison of Laboratory Frequency Standards.[a] [22]

Oscillator Type	Rubidium	Cesium Beam	GPSDO
Frequency offset with respect to UTC(NIST) (1 day average)	5×10^{-9} to 5×10^{-12}	1×10^{-12} to 5×10^{-14}	1×10^{-12} to 5×10^{-14}
Stability at 1 second	5×10^{-11} to 5×10^{-12}	5×10^{-11} to 5×10^{-12}	1×10^{-10} to 1×10^{-12}
Stability at 1 day	5×10^{-12}	8×10^{-14} to 2×10^{-14}	8×10^{-13} to 5×10^{-14}
Aging/year	$< 1 \times 10^{-10}$ to 5×10^{-10}	None, by definition. However, cesium oscillators have long-term frequency drift, typically measured in parts in 10^{17} over the course of a day.	None, the output is a steered frequency that is corrected for aging and drift.
Phase noise (dbc/Hz, 10 Hz from carrier)	-90 to -130	-130 to -136	-90 to -140
Life expectancy	> 15 years	5 to 20 years 10 years is typical	> 15 years
Produces an on-time pulse without being synchronized to another source	No	No	Yes
Produce frequency accurate to within $\pm 1 \times 10^{-11}$ for 24 hours or longer	Yes, with periodic Adjustment	Yes	Yes
Cost	$2,000 to $10,000	$30,000 to $55,000	$3,000 to $15,000

[a]This table compares laboratory grade frequency standards. However, low-cost quartz- and miniature Rb-based GPSDOs are available for under $1000.

than 100 ns during a 64 ms dwell period. The attained synchronisation was reported to be adequate for this radar. However, it should be noted that the S/A dithering was still active at the time. Much better synchronisation is achievable today using modern multi-channel GPS timing receivers with the S/A dithering switched off.

The Manastash Ridge Radar [85], an FM-based passive bistatic radar, used OCXO-based GPSDOs to synchronise the receiving and reference nodes across a baseline of roughly 100 km. Its carrier was at approximately 100 MHz. Moreover, this radar had the reference receiver located near the transmitter. The weak signal receiver was located about 100 km away where a mountain ridge helped to block the direct signal. The sampled data were recorded simultaneously at the reference and weak signal receivers and then processed centrally. A microwave link transported the data. The two receiving nodes were frequency synchronised to better than 10^{-11} and the GPS PPS time synchronisation had a mutual jitter of less than 100 ns.

The phase coherent bistatic radar described by Kesheng [14], and also discussed in Section 2.3.4, used multiple methods of synchronisation. Synchronisation across fibre and microwave were the primary methods, where Rb-based GPSDOs were used as fail-over backups in case both the other modes were unavailable. However, they did not report on the performance of the GPS synchronisation.

GPS synchronisation has also been used in bistatic high-frequency (HF) radar. The low carrier frequencies along with their low bandwidth and subsequently large range resolutions of such radars lowers the synchronisation requirements significantly. A bistatic high-frequency hybrid-sky surface wave radar (HFSSWR) system which is a combination of an HF skywave and HF surface wave radar was recently described [86]. This system is intended for long-range and wide-area monitoring and was demonstrated across a 780 km baseline. This FMCW system had a 19 MHz carrier and 40 kHz of bandwidth. It is not clear if OCXO- or Rb-based GPSDOs were used. It is also not clear how the time synchronisation mechanism works. However, it is assumed that the GPS PPS alone is adequate for the time synchronisation of this radar.

2.7   Conclusion

This chapter reviewed the literature related to the time, frequency, and phase synchronisation of bistatic, multistatic, and networked radar systems. Bi- and multistatic systems might offer various advantages over that of monostatic radar. However, the requirement of precise time, phase and frequency synchronisation remain one of the significant challenges.

Few publications discuss the time, frequency, and phase specification of bistatic radar from a systems perspective. However, it is often assumed that the time offset, frequency offset, and oscillator ageing is zero and that time and phase synchronisation are functions of frequency stability, $\sigma_y(\tau)$, only. This assumption is in the author's opinion not realistic where the relative frequency offset may often dominate in practice. Further, it remains difficult for the bistatic design engineer to quickly select a synchronisation technology based on the desired pulsed-Doppler performance.

The various synchronisation methods were compared including via RF and optic fibre cables, across RF or optical free-space links, and synchronisation using the direct radar emissions or clutter returns. The performance of each method is considered along with published examples where bistatic radars were synchronised using each technology.

GNSS time transfer is identified as uniquely suited to portable ground-based bistatic applications. It is cheap, simple, autonomous, passive, and does not require direct LOS. However, it remains susceptible to interference, jamming, and spoofing. The redundancy afforded by the multiple GNSS available today along with their improving performance keeps it an attractive option. Various methods of GNSS time transfer were considered. It is discussed how low-cost GPS timing receivers differ from the more common navigation receivers including the various timing error sources. Moreover, the GPS timing performance can be improved by several nanoseconds through calibration, accurately positional surveys, and antenna temperature stabilisation. The one-way method delivers long-term time accuracy and frequency stability of less than 20 ns and below 2×10^{-13} to UTC when averaged over 24 hours, respectively. However, it has a significant high-frequency noise component with a frequency uncertainty of a few parts in 10^{-9} averaged over one second, and a peak-to-peak time error of about 50 ns for 10-minute data point averages. Ground-based bistatic applications often have short baselines of tens up to a few hundred kilometres. Over such short baselines, the relative timing performance may be significantly better than what is quoted with respect to UTC. Moreover, over short baselines, the different GPS receivers view the same satellite constellation and are equally affected by the atmosphere. Results from Hambly and Clark [60] suggests that low-cost GPSDO smoothing should achieve a time error of $\pm 5\,$ns.

GPSDOs exploit the long-term stability of GPS, or other GNSS, to correct the time, phase and frequency drift effects caused by ambient changes and STALO imperfections. The result is an oscillator possessing both the short-term STALO performance and the long-term GPS stability. The published literature on GPSDO technology was reviewed. This review included various architectures and negative feedback strategies, time synchronisation mechanisms, and the synchronisation performance. A substantial opportunity for innovation lies within the GPSDO feedback control strategy which can either be PLL- or Kalman-based. The purpose of the feedback circuitry is to optimise the control loop bandwidth to maximise GPSDO performance. GPSDOs have excessive lock-in times ranging from tens of minutes to hours or days depending on the STALO stability. Such long lock-in periods are very limiting to portable bi- and multistatic applications which may require regular reacquisition of phase lock due to geometry changes. Few publications discuss quick-locking GPSDO feedback strategies. STALO hold-over temperature and ageing compensation, and time synchronisation mechanisms are also areas of interest. A rapid development environment for feedback and filtering algorithms that could be hardware tested directly is highly desirable. GPSDO performance is often specified for laboratory use. The long-term performance approaches that of Cesium with some Rb GPSDOs achieving $\sigma_y(\tau = 10^5) \leq 10^{-13}$. Weiss, Lisowiec, and Steplewski [25] postulated that a high-quality OCXO-based GPSDO is limited to a medium-term stability of $\sigma_y(\tau = 10^3) \approx 10^{-11}$, a long-term stability of $\sigma_y(\tau = 10^5) \approx 4 \times 10^{-13}$, and a peak-to-peak time error of below $40\,$ns. The results published in [22, 31, 81] appear to confirm these suggested limits. However, bistatic pulsed-Doppler applications are more concerned with the instantaneous Tx-Rx time and frequency accuracy at the time of target observation. Bistatic radar is also affected by the change in phase and frequency during the target observation, pulse integration, and Doppler integration. Moreover, the instantaneous GPSDO frequency may be orders of magnitude higher than that of the long-term averaged values. The instantaneous frequency offsets of OCXO-based GPSDOs ranged between $\pm 7 \times 10^{-11}$ and $\pm 2 \times 10^{-9}$ [81]. Further, Lombardi, Novick, and Zhang [22, 31] found that different commercial GPSDO models may have very different phase behaviours, and the manufacturer specified GPSDO performances of various models were found to be mere estimates of their actual performances. Well-calibrated GPSDOs with known behaviour is required to perform bistatic radar experiments.

Finally, the few published examples of GPS synchronised bistatic radar was reviewed. Much of what is published on the time and phase synchronisation requirements of radar is explicitly aimed at BSAR. Many of the publications also consider compensation techniques to correct for the imperfect synchronisation of BSAR systems. Much less is published on bistatic pulsed-Doppler synchronisation requirements or the effects of imperfect synchronisation on such systems. This is assumed to be because high-resolution SAR systems have such strict phase stability requirements. There are a few commercial BSAR systems in use today, but there are considerably fewer examples of commercial bistatic pulsed-Doppler systems reported in the literature. These commercial BSAR systems are often airborne or spaceborne with direct Tx-Rx LOS with synchronisation done via direct RF links. In such airborne applications, the problems of multipath and interference are much less severe than for ground-based systems. Also of note is that BSAR applications, in general, have shorter Tx-Rx baselines than typical ground-based bistatic pulsed-Doppler systems where the baselines may be tens to hundreds of kilometres. Another observation is that much of what is published on bistatic pulsed-Doppler synchronisation was in relation to experimental systems tested during the 1980's and 1990's. The lack of publications on pulsed-Doppler synchronisation may be ascribed to many possible reasons. For example, it could be that it is more problematic to time and phase synchronise over the larger baselines where LOS may sometimes not be available. It may also be that the tested experimental systems in the 80's and 90's were of lower performance and did not require such stringent synchronisation. GPS timing performance improved five-fold in the early 2000's when S/A was switched off. Thus, perhaps making it a more attractive solution to modern bistatic synchronisation. Regardless, the only recent examples of GPS synchronisation found was for VHF over the horizon radar. Further, the combined use of the sidelobe breakthrough phase to remove the inevitable GPSDO phase dynamics [27] appears to be a promising solution to low-cost but high-performance synchronisation. The performance of this technique must still be quantified and requires further investigation.

In the following chapter, the effect of imperfect synchronisation on bistatic pulsed-Doppler performance will be developed further. The goal is to make it easier for the designer to select an appropriate synchronisation technology based on the desired performance. Then, a set of GPSDOs are developed and calibrated with reduced lock-time times, accurate timing mechanisms, and a development environment to design and test new feedback algorithms rapidly. In the final chapter, these GPSDOs are used in combination with the sidelobe breakthrough phase to synchronise NetRAD to near monostatic levels.

Chapter 3

Bistatic Radar Synchronisation

This chapter expands on the current literature [3, 7–10, 20, 32] to better specify the synchronisation requirements of bistatic pulsed-Doppler radar. It starts by modelling an imperfectly synchronised bistatic Tx-Rx pair with the purpose to predict the effect of imperfect synchronisation on an otherwise ideal bistatic system. This model is based on the NetRAD system and used to directly relate oscillator performance to the bistatic range, Doppler and phase errors. The synchronisation requirements of bistatic non-coherent and coherent pulse integration are also derived. This model and error expressions are then used to define a set of synchronisation requirements for NetRAD based on the desired radar performance. Moreover, it is predicted that a low-cost quartz GPSDO is adequate to synchronise NetRAD. However, a bistatic system with independent STALOs has a much more stringent close-in phase noise requirement than a monostatic system if a similar SCV is required. Hence, LOS phase compensation is proposed to improve the bistatic phase performance cost-effectively across short baselines.

NetRAD is modelled as an imperfectly synchronised Tx-Rx pair. Moreover, it is assumed sufficient to consider the performance of single bistatic Tx-Rx pair and that it is scalable to the multistatic case. An existing and perhaps somewhat idealistic oscillator model [87, 88] was used. However, more complex models could substitute this model. Moreover, the total accumulated relative phase error is lumped together at the receiver. The first goal is to express the receiver IF signal as a function of transmitter-target-receiver time delay, bistatic doppler frequency, and imperfect Tx-Rx synchronisation. Then, this IF signal is sampled and expressed as a function of radar pulse number, analogue-to-digital converter (ADC) sample number, and an arbitrary future time when the first radar pulse is sampled. Finally, the individual phase, Doppler, and radar range error contributions by the heterodyne conversion, radar PRF and ADC sampling are derived. The total error contribution is calculated from the individual error contributions. It is shown that ADC sampling does not introduce a significant error over that of down-conversion for stable oscillators. A set of plots directly relating oscillator parameters to radar performance is presented. These plots can be used to quickly gauge the Tx-Rx synchronisation demands based on the required bistatic performance.

A constant frequency offset may often dominate the Tx-Rx synchronisation error for STALOs. Hence, it is appropriate to investigate the effect of a constant frequency offset, or a linear phase ramp, on pulse integration. Moreover, bistatic pulse integration is equivalent to the monostatic integration of a target travelling at a constant radial velocity. In non-coherent processing, this range advance results in a reduction in amplitude gain, a broadening of the main lobe of the target, and a range error. During coherent processing, there is a loss of coherent integration gain. Moreover, it is shown that a total phase creep of 95°, 160° and 266° result in an SNR loss of 1 dB, 3 dB and 10 dB, respectively.

Finally, the required synchronisation to achieve the desired NetRAD performance is specified. Moreover, the model and plots developed throughout the chapter are used to set the required time, frequency and phase synchronisation to achieve the desired bistatic range, Doppler and phase performance. Low-cost quartz GPSDOs are estimated to be adequate to synchronise NetRAD. However, the desired range accuracy must be relaxed from one-tenth of the range resolution to just within the range resolution or a range bin of 3 metres. Also, it appears impossible for low-cost GPSDOs to provide the required ultra-low close-in phase noise to ensure an SCV equivalent to that of monostatic radar. Hence, a LOS phase compensation technique is proposed where the breakthrough along a static baseline is used to remove GPSDO induced phase dynamics. This technique is to be used complementary to low-cost GPSDOs to improve the bistatic phase performance cost-effectively across short baselines. A set of three low-cost one-way GPSDOs are developed in Chapter 4. These GPSDOs are calibrated in Chapter 5 using the low-cost DMTD designed in Appendix C and then later used to synchronised NetRAD in Chapter 6. These bistatic measurements are analysed to verify the predicted bistatic performance of this chapter.

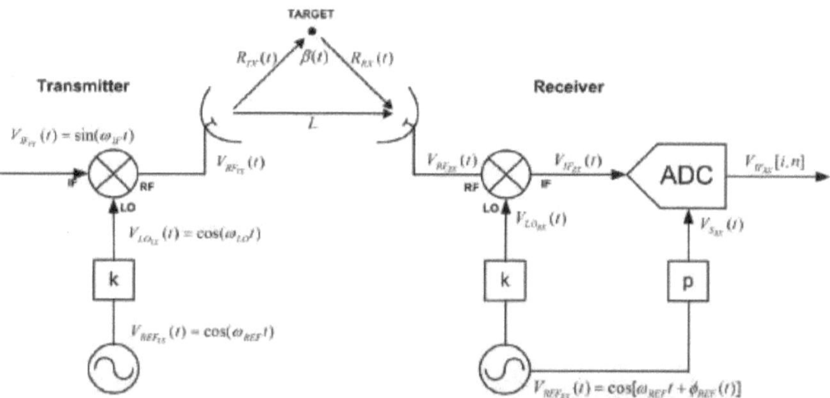

Figure 3.1: Model of NetRAD Bistatic Synchronisation.

3.1 Model of Bistatic Radar Synchronisation

This section models an imperfectly synchronised bistatic Tx-Rx pair of an otherwise ideal bistatic NetRAD configuration.

NetRAD is a coherent tri-node pulsed-Doppler network radar. The carrier frequency is at S-band (2.4 GHz) with a bandwidth of 50 MHz. The NetRAD transmitter is a fully monostatic node that can transmit as well as receive. The NetRAD receiving nodes are passive and receive only. The IF is at baseband where it coherently samples the real-valued signal. A software Hilbert transform converts these real-valued samples to a complex signal. Each node has an independent reference oscillator from which it derives the LO and sampling frequencies, and radar PRF. Refer to Table 3.1 for the basic NetRAD [12, 13] specifications.

The NetRAD synchronisation is modelled as a transmitter up-conversation stage and a receiver down-conversion and sampling stage. The transmitter's reference oscillator is assumed to be ideal, whereas the total relative time-dependent Tx-Rx phase deviation is lumped together within the receiver's reference oscillator. This model considers the effects of imperfect synchronisation on heterodyne down-conversion, the radar PRF, and ADC sampling.

The first goal is to express the receiver IF signal as a function of transmitter-target-receiver time delay, bistatic doppler frequency, and imperfect Tx-Rx synchronisation. Then, this IF signal is sampled and expressed as a function of radar pulse number, ADC sample number, and an arbitrary future time at which the first radar pulse is sampled. Finally, the individual phase, Doppler, and radar range error contributions from the heterodyne conversion, radar PRF and ADC sampling are derived. From these individual error contributions, the total error contribution is calculated.

For the remainder of this section, refer to the NetRAD bistatic synchronisation model depicted in Figure 3.1.

3.1.1 Radar Geometry

The moving target, a non-fluctuating perfect point scatterer, is located at a distance $R_{TX}(t)$ from the transmitter, and a distance $R_{RX}(t)$ from the receiver, where the distance L is the bistatic baseline (Tx-Rx distance). The transmitter and receiver are assumed to be on static platforms. The bistatic radar range [3] is defined as the sum of the transmitter-target distance and the target-receiver distance minus the bistatic baseline such that

$$R_B(t) = R_{TX}(t) + R_{RX}(t) - L. \qquad \text{[m]} \quad (3.1)$$

The bistatic doppler frequency [3] is dependent on the rate of change of the sum of the transmitter-target and target-receiver-distances and is defined as

$$\omega_{D_B}(t) = \frac{\omega_{RF}}{c} \frac{d}{dt}[R_{TX}(t) + R_{RX}(t)], \qquad \text{[rad/s]} \quad (3.2)$$

where c is the speed of light, ω_{RF} is the radar carrier frequency.

Note that for the quasi-monostatic case, where the transmitter and receiver are co-located, $L = 0$, and the radar range and Doppler in (3.1) and (3.2) reduce to their monostatic versions

$$R_M(t) = R_{TX}(t) = R_{RX}(t) \qquad \text{[m]} \quad (3.3)$$

and

$$\omega_{D_M}(t) = \frac{2\omega_{RF}}{c} \frac{d}{dt}[R_M(t)], \qquad \text{[rad/s]} \quad (3.4)$$

respectively.

3.1.2 Transmitter

The transmitter section consists of a heterodyne up-conversion stage. The transmitted IF signal is mixed with the LO signal and high-pass filtered to produce the RF frequency. The reference oscillator is multiplied by a factor k to generate the mixer LO frequency. Here it is sought to model only the deterministic and long-term random oscillator effects. Hence, it is sufficient to assume simple frequency multiplication and set all signal amplitudes to unity. The radar operates at a nominal reference oscillator frequency, ω_{REF}, a nominal LO frequency, ω_{LO}, a nominal IF frequency, ω_{IF}, and a nominal RF frequency, ω_{RF}. The transmitter's frequency reference signal, $V_{REF_{TX}}(t)$, mixer LO signal, $V_{LO_{TX}}(t)$, IF signal, $V_{IF_{TX}}(t)$, are assumed to be perfect sinusoids, and the RF signal, $V_{RF_{TX}}(t)$ is

$$V_{RF_{TX}}(t) = \sin(\omega_{RF}t). \tag{3.5}$$

where $\omega_{LO} = k\omega_{REF}$ and $\omega_{RF} = \omega_{LO} + \omega_{IF}$.

3.1.3 Receiver

The transmitted pulse is reflected off the target and received by the receiver. The received RF signal, $V_{RF_{RX}}(t)$, is a delayed version of the transmitted signal, $V_{RF_{TX}}(t)$, with an additional bistatic doppler frequency offset, $\omega_D(t)$, such that

$$V_{RF_{RX}}(t) = \sin[(\omega_{RF} + \omega_D(t))t - \omega_{RF}\tau_B(t)], \tag{3.6}$$

where

$$\tau_B(t) = \frac{[R_{TX}(t) + R_{RX}(t)]}{c}. \qquad [\text{s}] \quad (3.7)$$

The bistatic time delay, $\tau_B(t)$, is the time it took for the reflected signal to reach the receiver from the moment it was transmitted. When the baseline L is known, the radar calculates range by measuring $\tau_B(t)$.

The receiver section consists of a heterodyne down-conversion stage and an ADC sampling stage. The received RF signal is mixed with the mixer LO frequency and low-pass filtered to produce the receiver IF signal. This IF signal is then sampled at the radar PRF and ADC sampling frequency. As stated earlier, the total relative Tx-Rx synchronisation error is included in the receiver's reference oscillator. Similar to (A.1), the total relative time dependant Tx-Rx phase deviation is lumped together in the receiver's reference oscillator signal, $V_{LO_{REF}}(t)$, as the time-dependent phase residual, $\phi_{REF}(t)$, such that

$$V_{REF_{RX}}(t) = \cos[\omega_{REF}t + \phi_{REF}(t)], \qquad (3.8)$$

where $\phi_{REF}(t) = \omega_{REF}x_\epsilon(t)$.

The model[1][2] from (A.6) is now used to give the total cumulative time deviation as

$$x_\epsilon(t) = x_o + y_o t + \Delta T K_T t + \frac{1}{2}Dt^2 + \sigma_x(t). \qquad [\text{s}] \quad (3.9)$$

The Tx-Rx synchronisation is now expressed as a function of, x_o, the initial relative time offset, $y_{o,}$, the initial relative frequency offset, ΔT, the total change in temperature, K_T, the oscillator temperature coefficient, D, the reference oscillator's relative frequency drift, and, $\sigma_x(t)$, the time Allan deviation representing the relative random time-dependent fluctuations.

Identical to the transmitter, the receiver's reference oscillator is multiplied by a factor k to generate the mixer LO such that

$$V_{LO_{RX}}(t) = \cos(\omega_{LO}t + \phi_{LO}(t)). \qquad (3.10)$$

The receiver's IF signal can now be expressed as a function of transmitter-target-receiver time delay, $\tau_B(\text{t})$, bistatic doppler frequency, $\omega_{D_B}(t)$, and relative Tx-Rx synchronisation, $\phi_{LO}(t)$, as

$$V_{IF_{RX}}(t) = \sin[(\omega_{IF} + \omega_{D_B}(t))t - \omega_{LO}\tau_B(t) - \phi_{LO}(t)]. \qquad (3.11)$$

If there is a direct Tx-Rx LOS, there may be a direct transmission through the antenna sidelobes referred to as the breakthrough. For a static baseline, the breakthrough has a constant time delay, τ_L, and no doppler offset. It follows that the breakthrough produces the IF signal

$$V_{IF_{RXL}}(t) = \sin[\omega_{IF}t - \omega_{LO}\tau_L(t) - \phi_{LO}(t)]. \qquad (3.12)$$

---

[1]Other more complex oscillator models may substitute this model.

[2]It is conventional to specify oscillator ageing as per day or month. This model assumes a linear ageing rate in units of per second. There are 86400 seconds in a day which is roughly 10^5 on a logarithmic plot.

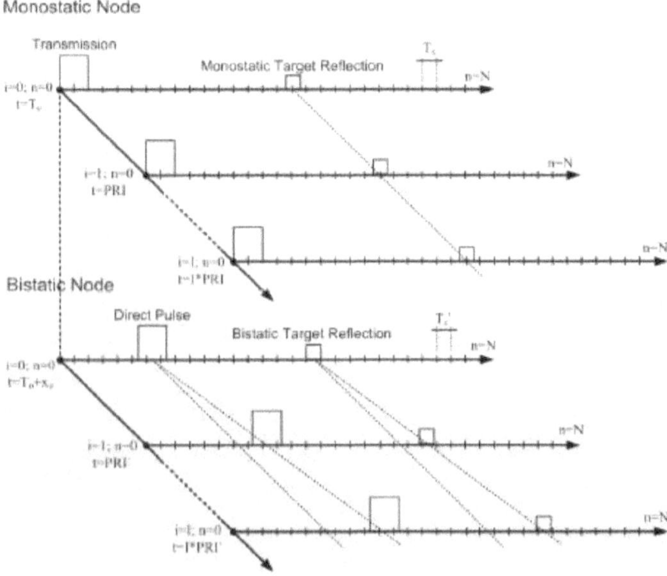

Figure 3.2: NetRAD Bistatic Timing Diagram. A monostatic transceiver has a common time base with no Tx-Rx PRF and ADC sampling discrepancy. However, in a non-synchronous bistatic receiver, the sampling timebases are offset and drifting which distorts its perception of the target.

For a known baseline, L, the phase $\phi_{LO}(t)$ is the only unknown in (3.12). Thus, the breakthrough can potentially be used to correct the Tx-Rx synchronisation in post-processing. Refer to Section 3.3 where LOS phase compensation is discussed further.

3.1.4 Sampling: PRF and ADC

At the sampling stage, the received IF frequency is sampled by both the slower PRF, and the faster sampling frequency. Refer to the radar slow-fast timing diagram in Figure 3.2. For each measurement, NetRAD records a set I pulses each containing N samples which may start at an arbitrary future time, T_o. The pulses are spaced apart by the nominal pulse repetition interval (PRI), where PRI $= 1/$PRF, and the samples are spaced apart by the nominal sampling period, Ts, where $Ts = 1/v_s$.

A monostatic transceiver has a common time base with no Tx-Rx PRF and ADC sampling discrepancy. However, in a non-synchronous bistatic receiver, the sampling timebases are offset and drifting which distorts its perception of the target.

In Figure 3.1, the receiver's reference oscillator is multiplied by a factor p to generate the ADC sampling timebase, $V_{S_{RX}}(t)$, such that

$$V_{S_{RX}}(t) = \cos[(\omega_S + \phi_S(t)], \tag{3.13}$$

where $\omega_S = p\omega_{REF}$ and the nominal sampling period is $T_s = 2\pi/\omega_s$.

For a bistatic measurement which started at an arbitrary time, $t = T_o$, the PRI timebase is modulated due to imperfect reference synchronisation such that

$$i\text{PRI}' \quad = \quad i\text{PRI} + x_\epsilon(i\text{PRI} + T_o), \qquad \text{[s]} \quad (3.14)$$

for $0 \leq i \leq I - 1$, where PRI$'$ represents the bistatic PRI following the ith pulse. The ith PRI$'$ is the time difference between pulses $(i + 1)$ and i such that

$$\text{PRI}'[i] = \text{PRI} + x_\epsilon[(i + 1)\text{PRI} + T_o] - x_\epsilon[i\text{PRI} + T_o], \qquad \text{[s]} \quad (3.15)$$

for $0 \leq i \leq I - 2$. It follows that the radar PRF averaged over the ith PRI$'$ is

$$\text{PRF}'[i] = \frac{1}{\text{PRI}'[i]}. \qquad \text{[Hz]} \quad (3.16)$$

If one now assumes the ADC starts to sample immediately at the start of each PRI, and that the ADC timebase is also modulated due to imperfect reference synchronisation one have

$$i\text{PRI}' + nT_s' \quad = \quad (i\text{PRI} + nT_s) + x_\epsilon(i\text{PRI} + nT_s + T_o). \qquad \text{[s]} \quad (3.17)$$

for $0 \leq i \leq I - 1$ and $0 \leq n \leq N - 1$, where T_s' represents the bistatic sampling period following the nth ADC sample. The nth sampling period, T_s', of the ith radar pulse is then the time difference between samples $(n + 1)$ and n of pulse i such that

$$T_s'[i, n] = T_s + x_\epsilon[i\text{PRI} + (n + 1)T_s + T_o] - x_\epsilon[i\text{PRI} + nT_s + T_o], \qquad \text{[s]} \quad (3.18)$$

for $0 \leq i \leq I - 2$ and $0 \leq n \leq N - 2$. It follows that the ADC sampling frequency averaged over the nth sample of the ith pulse is

$$v'_s[i, n] = \frac{1}{T'_s[i, n]}. \qquad \text{[Hz]} \qquad (3.19)$$

Finally, with the nominal PRF and sampling frequency known, the bistatic pulsed and sampled IF signal can be written as a function of pulse number, i, sample number, n, and the start of transmission time, T_o, such that

$$V_{IF_{RX}}[i, n, T_o] = V_{IF_{RX}}[(i\text{PRI} + nT_s) + x_\epsilon(i\text{PRI} + nT_s + T_o)]. \qquad (3.20)$$

This achieves the goal of expressing the sampled bistatic IF signal in terms of Tx-Rx synchronisation.

3.2 Bistatic Phase, Doppler and Range Errors

The previous sections modelled a bistatic Tx-Rx pair with imperfect synchronisation. This section derives error expressions for the bistatic phase, Doppler and range as a function of oscillator synchronisation. These errors are plotted for a range of initial Tx-Rx frequency offsets, temperature offset, and drift rates. These plots can be used to quickly gauge the Tx-Rx synchronisation demands based on the required bistatic performance.

First, the combined error introduced by down-conversion only is considered. It is assumed that both the target range and Doppler frequency is constant during the radar observation such that the phase of the IF signal in (3.11) reduces to

$$\Phi'_{IF}(t) \;=\; \omega_{IF}t + \omega_D t - \omega_{LO}\tau_B - \omega_{LO}x_\epsilon(t). \qquad \text{[rad]} \qquad (3.21)$$

In the case where the Tx-Rx pair is perfectly synchronous, the phase of the IF signal (3.11) becomes

$$\Phi_{IF}(t) \;=\; \omega_{IF}t + \omega_D t - \omega_{LO}\tau_B. \qquad \text{[rad]} \qquad (3.22)$$

The instantaneous phase error, introduced by down-conversion only, is now defined as the difference between the ideal and non-synchronous phases to give

$$\Delta\Phi_{IF}(t) \quad = \quad \Phi_{IF}(t) - \Phi'_{IF}(t) \qquad\qquad (3.23)$$

$$= \quad \omega_{LO}x_\epsilon(t), \qquad\qquad [\text{rad}] \quad (3.24)$$

and from (3.23) in (A.2) the instantaneous frequency error due to down-conversion is

$$\Delta v_{IF}(t) \quad = \quad y_\epsilon(t)v_{LO}, \qquad\qquad [\text{Hz}] \quad (3.25)$$

where

$$y_\epsilon(t) = y_o + Dt + \Delta T K_T + \sigma_y(t). \qquad\qquad [\ ] \quad (3.26)$$

Next, the error introduced by both the down-conversion and sampling of the IF signal is considered. In the ideal case, where the Tx-Rx pair is perfectly synchronous, an ideal IF signal is sampled by a perfect timebase. From (3.22) the phase of a perfectly synchronous sampled IF signal yields

$$\Phi_{IF}[i, n, T_o] \quad = \quad \omega_{IF}T + \omega_D T - \omega_{LO}\tau_B, \qquad [\text{rad}] \quad (3.27)$$

where $T = i\text{PRI} + nT_s$.

However, where the Tx-Rx pair is non-synchronous the phase of the non-synchronous sampled IF signal from (3.21) in (3.14) is

$$\Phi'_{IF}[i, n, T_o] = \omega_{IF}[T + x_\epsilon(T)] + \omega_D[T + x_\epsilon(T)]$$
$$- \omega_{LO}\tau_B - \omega_{LO}x_\epsilon[T + x_\epsilon(T)], \qquad [\text{rad}] \quad (3.28)$$

where it is assumed that $T_o = 0$. Note that the non-synchronous sampling introduces a time dependent error into each of the terms in (3.21). The error introduced by down-conversion now contains a more complex error term $x_\epsilon[T + x_\epsilon(T)]$.

Again, the instantaneous phase error introduced by down-conversion and sampling is defined as the difference between the synchronous and non-synchronous sampled phases such that

$$\Delta\Phi_{IF}[i,n,T_o] \quad = \quad \Phi_{IF}[i,n,T_o] - \Phi'_{IF}[i,n,T_o] \tag{3.29}$$

$$= \quad -\omega_{IF}x_\epsilon(T) - \omega_D x_\epsilon(T) + \omega_{LO}x_\epsilon[T + x_\epsilon(T)]. \tag{3.30}$$

Equation 3.29 expresses the combined phase error introduced by non-synchronous down-conversion and sampling as the sum of three terms. The factor $x_\epsilon[T + x_\epsilon(T)]$ is a complex polynomial containing many terms. However, for most STALOs, the values of y_o, D, and σ_x are typically on the order of 10^{-7} down to 10^{-12} and possibly smaller. If one further assumes $x_o \ll 0$, the higher-order terms become negligibly small and $x_\epsilon[T + x_\epsilon(T)] \approx x_\epsilon(T)$. Thus, for most stable oscillators, sampling does not introduce a significant frequency error in addition to that of down-conversion. Also, for most radar carrier, Doppler and IF frequencies it can be assumed that $\omega_{LO} \ll \omega_{IF}$ and $\omega_{LO} \ll \omega_D$ where (3.29) reduces to

$$\Delta\Phi_{IF}[i,n,T_o] \quad \approx \quad (\omega_{LO} - \omega_D - \omega_{IF})x_\epsilon(T) \tag{3.31}$$

$$\approx \quad \omega_{LO}x_\epsilon(T). \qquad\qquad [\text{rad}] \tag{3.32}$$

Figure 3.3 compares the hold-over bistatic carrier phase error for typical Quartz, Rubidium and Cesium standards. It is seen that Quartz and Rubidium standards are only useful in coherent bistatic applications when the Tx-Rx frequencies are regularly synchronised. Free-running Quartz standards are unlikely to be able to provide adequate hold-over performance for coherent bistatic radar.

The instantanous frequency error introduced by non-synchronous sampling now yields from from (3.31) in (A.2)

$$\Delta v_{IF}[i,n,T_o] \approx v_{LO}y_\epsilon(T). \qquad\qquad [\text{Hz}] \tag{3.33}$$

Refer to the plots in Figures 3.4 and 3.5 for the Doppler error due to Tx-Rx frequency offset and daily frequency drift, respectively. These plots assume baseband operation where $\omega_{IF} \approx 0$ such that $\omega_{RF} \approx \omega_{LO}$.

As an example, if one has a carrier frequency of 2.4 GHz, and it is required that the Doppler error must stay within 1 Hz. Then, from Figure 3.4 it is seen that

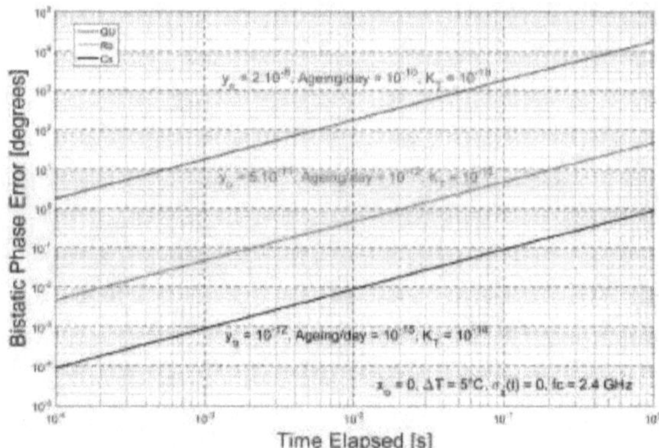

(a) Non-Synchronised: Frequency offset, y_o, is set to typical factory shipped values. After one second of hold-over, the Quartz, Rubidium and Cesium standards drift by 10,000°, 50° and 1°, respectively

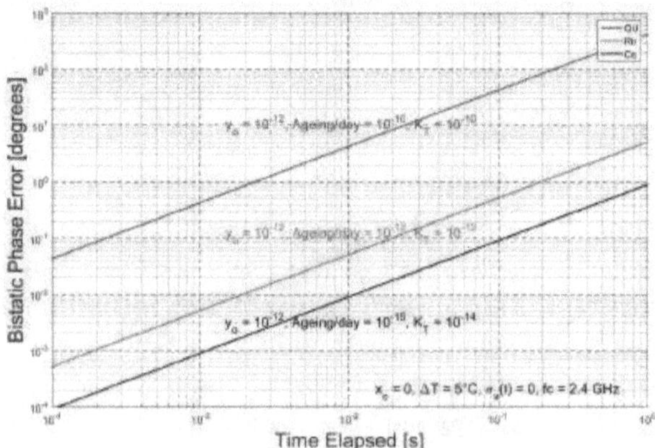

(b) Synchronised: The standards are initially synchronised to $y_o = 10^{-12}$. After one second of hold-over, the Quartz, Rubidium and Cesium standards drift by 400°, 4° and 1°, respectively

Figure 3.3: Bistatic Hold-Over Phase Error versus Time Elapsed. The bistatic hold-over phase error for Quartz, Rubidium and Cesium are compared. Typical STALO ageing and temperature coefficients, a Tx-Rx temperature gradient of 5°C, and a carrier frequency of 2.4 GHz are used. For simplicity, the random effects, $\sigma_x(t)$, are assumed zero.

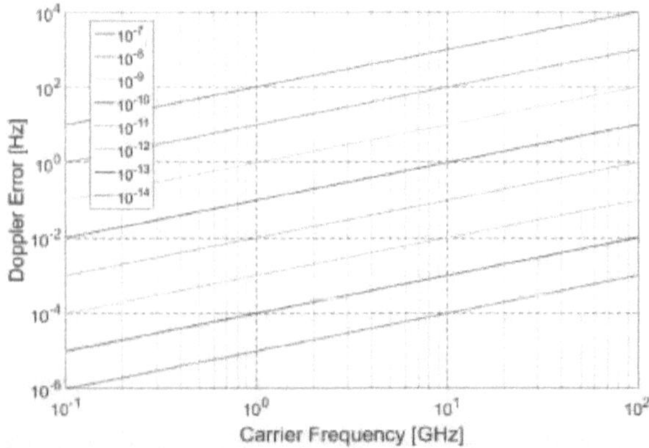

Figure 3.4: Bistatic Doppler Error versus Carrier Frequency at Various Frequency Offsets. ($D = 0$, $\sigma_y = 0$) For a Doppler error less than 1 Hz at a carrier frequency of 2.4 GHz, the frequency offset must be below 5×10^{-10}.

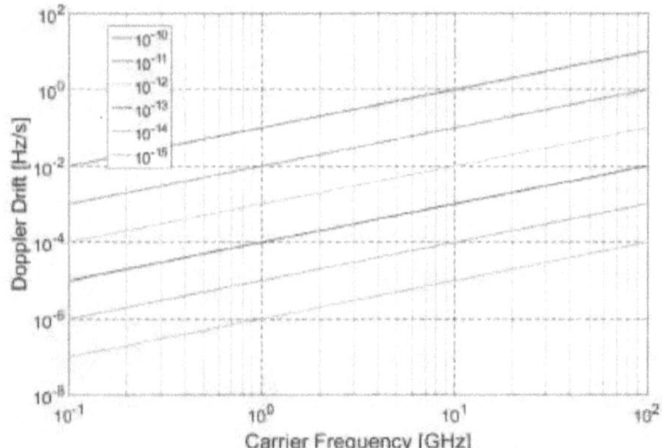

Figure 3.5: Bistatic Doppler Drift versus Carrier Frequency at Various Frequency Drift Rates. ($y_o = 0$, $K_T = 0$, $\sigma_y = 0$) Drift is specified in fractional frequency per day. If a 2.4 GHz system is re-synchronised every 1000 seconds and requires a Doppler error of less than 1 Hz, the Doppler drift must be better than $1/1000 = 10^{-3}$ Hz/s requiring a Rubidium level ageing rate of less than 5×10^{-13} per day.

a frequency offset of better than 5×10^{-10} is required. For a Tx-Rx pair which is re-synchronised once every 1000 seconds, the Doppler drift must be better than $1/1000 = 10^{-3}$ Hz/s. Hence, from Figure 3.5 it is required that the relative frequency drift is less than 5×10^{-13} per day. Thus, during hold-over, such an application requires Rubidium level ageing.

Finally, the error in the measured bistatic time delay is considered. The radar measures time delay with its sampling timebase, and a non-synchronous timebase introduces a time dependent error such that

$$\tau'_B[i, n, T_o] \quad = \quad \tau_B[i, n, T_o] + x_\epsilon(i\mathrm{PRI} + nT_s + T_o), \qquad [\mathrm{s}] \quad (3.34)$$

where τ_B is the ideal time delay and τ'_B is the time delay measured using the non-synchronous timebase. From (3.34), the error between the ideally and non-synchronously measured time delay is

$$\Delta\tau_B[i, n, T_o] \quad = \quad -x_\epsilon(i\mathrm{PRI} + nT_s + T_o), \qquad\qquad (3.35)$$
$$[\mathrm{s}] \quad (3.36)$$

and from (3.7) the error in the bistatic range becomes

$$\Delta R_B[i, n, T_o] \quad = \quad -cx_\epsilon(i\mathrm{PRI} + nT_s + T_o). \qquad [\mathrm{m}] \quad (3.37)$$

Note that the radar range error due to Tx-Rx synchronisation is only a function of $x_\epsilon(t)$ and the total elapsed time. It is independent of any radar parameter.

Refer to the plots in Figures 3.6 to 3.8 for the bistatic range error due to a time offset, a frequency offset, and frequency drift, respectively.

Further, it may be possible to measure the Tx-Rx synchronisation error at regular intervals using an RF link or similar. If the time and frequency offsets and drift rate are known, the PRI and ADC sampling intervals may be synchronised in software. To this end, the PRI and sampling errors are expressed as functions of the clock synchronisation. Moreover, the time error of the ith PRI is given by the difference between the ideal and non-synchronous PRI from (3.15) as

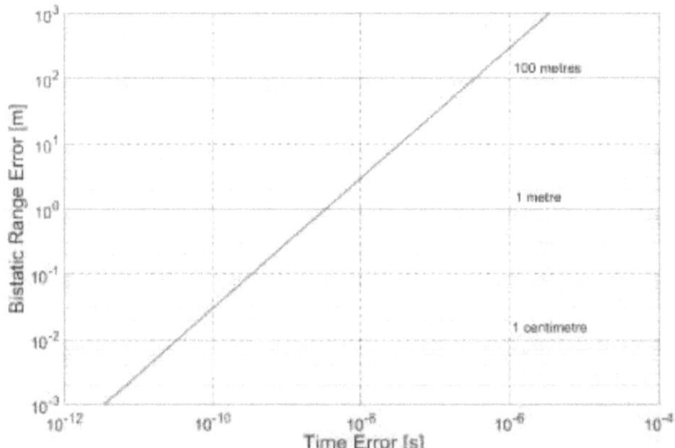

Figure 3.6: Bistatic Range Error due to Time Offset. ($y_o = 0$, $K_T = 0$, $D = 0$, $\sigma_x = 0$) Time offsets of 3.3 ps, 3.3 ns, and 33 ns produce bistatic range errors of 1 cm, 1 m, and 100 m, respectively.

$$\Delta\mathrm{PRI}'[i, T_o] = \mathrm{PRI} - \mathrm{PRI}'[i, T_o] \tag{3.38}$$

$$= x_\epsilon[i\mathrm{PRI} + T_o] - x_\epsilon[(i+1)\mathrm{PRI} + T_o]. \quad [\mathrm{s}] \tag{3.39}$$

Now the frequency error of the bistatic PRF averaged over the ith pulse yields

$$\Delta\mathrm{PRF}'[i] = \frac{1}{\Delta\mathrm{PRI}'[i]}. \quad [\mathrm{Hz}] \tag{3.40}$$

Similarly, the time error of the nth ADC sample is given by difference between the ideal and non-synchronous sampling periods from (3.18) is

$$\Delta T'_s[i, n] = T_s - T'_s[i, n, T_o] \tag{3.41}$$

$$= x_\epsilon[i\mathrm{PRI} + nT_s + T_o] - x_\epsilon[i\mathrm{PRI} + (n+1)T_s + T_o], \tag{3.42}$$

where the frequency error of the bistatic sampling frequency averaged over the

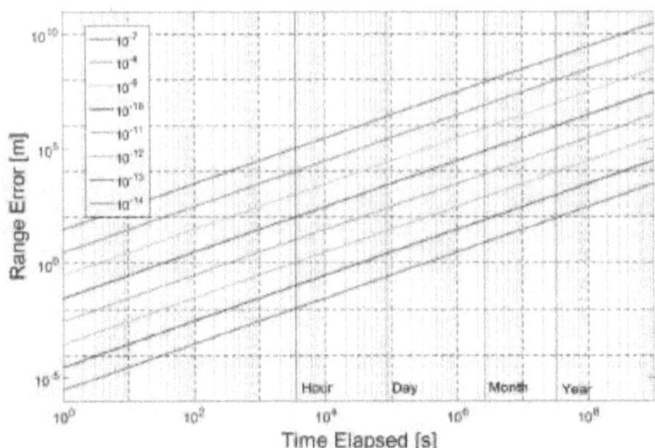

Figure 3.7: Bistatic Range Error versus Time for Various Frequency Offsets. ($x_o = 0$, $D = 0$, $\sigma_x = 0$) For $y_o = 10^{-12}$ the bistatic range error reaches one metre only after one hour.

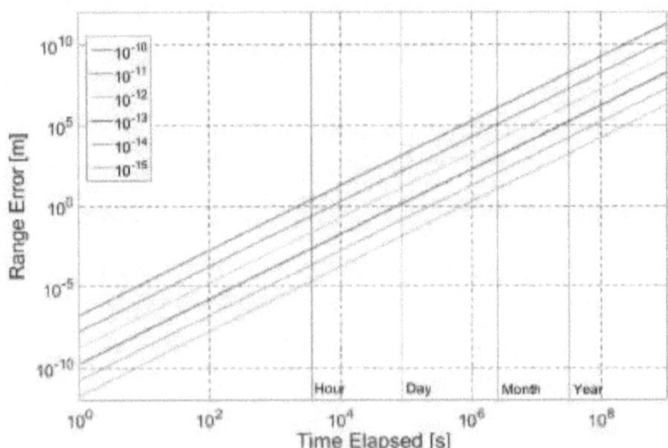

Figure 3.8: Bistatic Range Error versus Time for Various Frequency Drift Rates. ($x_o = 0$, $K_T = 0$, $y_o = 0$, $\sigma_x = 0$) Drift is specified in fractional frequency per day. Quartz (10^{-10}) drifts by one metre in less than an hour. However, a good Rubidium (10^{-13}) may drift less than one metre in a day.

n^{th} sample of the ith pulse is

$$\Delta v'_s[i, n] = \frac{1}{\Delta T'_s[i, n]}. \qquad \text{[Hz]} \quad (3.43)$$

Thus, the goal of deriving error expressions for the IF phase, target Doppler, target range, and the PRI and ADC sampling intervals as functions of bistatic synchronisation is achieved.

3.3 LOS Phase Compensation

This section discusses LOS phase compensation[1] as a method to suppress GPSDO induced target phase dynamics. LOS phase compensation has the potential to offer high-performance phase coherence at both a low monetary and computational cost. For a known baseline, L, and for ideal wave propagation, it follows from (3.12) that the perceived IF phase is $\phi_{LO}(t)$ where $\phi_{LO}(t)$ is a function of Tx-Rx synchronisation only. Thus, the breakthrough can be used to measure the Tx-Rx time (phase), frequency and frequency drift. However, real applications are subject to multipath, interference, propagation effects and often require node cooperation [3]. Bistatic time and frequency synchronisation schemes using the transmitted radar emissions have been demonstrated in the literature since the early 1980's [7, 9, 15, 16] and used the direct signal as well as common target or clutter reflections. However, little is published on the real-world performance of sidelobe phase synchronisation.

In GPSDO-synchronised bistatic systems, the time and frequency are already synchronised making it possible to record both the breakthrough and bistatic target within a single pulse. Then, all the samples in each pulse can be time-shifted equally to null the breakthrough phase. Relatively little computation is required when using the time shift property of the Fourier transform

$$\mathcal{F}[x(t \pm t_o)] = X(j\omega)e^{\pm j\omega t_o}. \qquad (3.44)$$

This technique is demonstrated in Chapter 6 where breakthrough phase compensation is performed on real GPSDO synchronised bistatic pulsed-Doppler data

---

[1] Al-Ashwal first demonstrated how this technique could improve the Doppler performance of the GPSDO synchronised NetRAD. [27]

across baselines of a few kilometres. The achieved phase, frequency, and Doppler performance is reported.

3.4 Bistatic Pulse Integration

This section investigates the effect of imperfect frequency synchronisation on both non-coherent and coherent bistatic integration.

It is common for a pulsed radar to integrate pulses to improve the target SNR in the presence of noise. In most applications, the total integration period spans at most a few seconds but often less than one second. It is sufficient to assume a static Tx-Rx frequency offset with negligible drift for most STALOs during such typical integration periods. The bistatic frequency offset induces a linear range (phase) advance from pulse to pulse. Hence, bistatic pulse integration is equivalent to the monostatic integration of a target travelling at a constant radial velocity. This is known as target range migration. The subject of range migration and its mitigation for monostatic radar is well-studied in the literature [89]. Moreover, the author believes that much of the current literature on range migration may be directly applicable to the case of a constant bistatic frequency offset and possibly frequency acceleration. However, this is left to future work to investigate this relationship.

The goal of this section is not to make an in-depth study of bistatic pulse integration. The purpose is to enable the designer to specify the level of bistatic synchronisation based on the desired integration performance. This is done in a fairly rudimentary manner. Moreover, for non-coherent integration, the achieved integration amplitude gain, the target's perceived range, and the width of the targets main lobe are derived as functions of total range creep. The loss in bistatic non-coherent SNR due to range creep is not quantified and left for future work. For coherent integration, the integration SNR loss is derived as a function of the total phase creep. The validity of these derivations are later tested in Section 6.3.9 and Section 6.3.10 using real GPSDO synchronised bistatic pulsed-Doppler data.

3.4.1 Non-coherent Integration

This section relates bistatic frequency offset to non-coherent integration performance. Moreover, a constant frequency offset causes the bistatic target range to advance linearly from pulse to pulse. This bistatic range creep is plotted for various

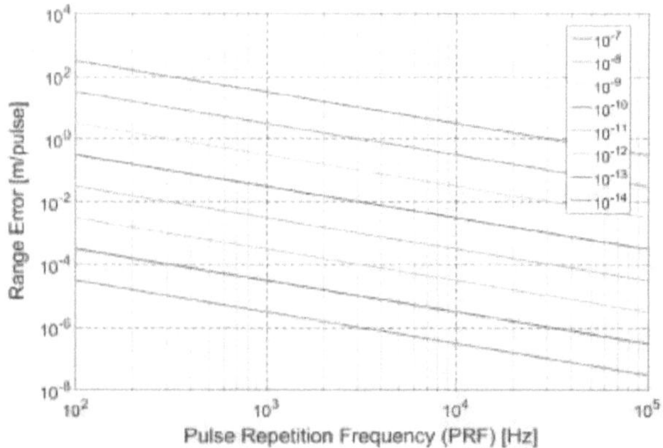

Figure 3.9: Pulse-to-Pulse Bistatic Range Error versus PRF for Various Frequency Offsets. A frequency offset of approximately 5.10^{-9} results in a range creep of about $1\,\mathrm{m/pulse}$ for PRF $= 1\,\mathrm{kHz}$.

Tx-Rx frequency offsets in Figure 3.9.

One may start by modelling the compressed radar target amplitude $A_{target}(a, b)$ as the squared sinc function

$$A_{tgt}(a, b) = \left(\frac{\sin[X(a, b)]}{X(a, b)} \right)^2,\tag{3.45}$$

where

$$X(a, b) = \frac{2\pi}{2.26}(a - b),\tag{3.46}$$

where a is the dependent variable for radar range, and b is the constant range of the target. Both a and b are normalised to the main lobe's $3\,\mathrm{dB}$ width which is purposely scaled to unity.

One can now do amplitude, or non-coherent, integration, by summing N successive pulses together. However, it is further assumed that two bistatic nodes have a constant relative frequency offset. This effect can be expressed as

$$A_{tgtN}(a, b) = \frac{1}{N} \sum_{i=1}^{N} \left(\frac{\sin[X_N(a, b)]}{X_N(a, b)} \right)^2,\tag{3.47}$$

where

$$X_N(a,b) = \frac{2\pi}{2.26}\left(a - \frac{ib}{N}\right),\qquad(3.48)$$

where $A_{tgtN}(a,b)$ is the normalised integrated target amplitude, b is the total normalised distance through which the target as has moved during the integration period, and N is the total number of pulses integrated.

The normalised non-coherent amplitude gain as a function of total target creep distance, b, is given by the maximum value of integrated target amplitude

$$A_{nc}(b) = \max\left[A_{tgtN}(a,b)\right].\qquad(3.49)$$

The normalised non-coherent amplitude gain, $A_{nc}(b)$, versus range creep, b, is plotted in Figure 3.10. The normalised non-coherent range error versus range creep is plotted in Figure 3.11, and the normalised 3 dB main lobe width of the target is plotted versus range creep, b, in Figure 3.12. These plots quantify what is intuitively expected. The amplitude gain diminishes, the target's main lobe widens, and the perceived target range advances with increasing range creep.

For example, let us assume it is required that a target creeps by no more than a 1 m range bin during the non-coherent integration of 100 pulses. From Figure 3.10, the loss in amplitude gain is less than 20% and from Figure 3.11 the range error is less than half a range bin. Also, from Figure 3.9 a frequency offset better than about $5.10^{-9}/100 = 5.10^{-11}$ is required at a PRF of 1 kHz.

Thus, synchronisation can now be specified based on non-coherent integration performance. It is also desirable to quantify the loss of non-coherent SNR due to range creep. However, the non-linear nature of the envelope detection process makes this a more difficult problem to solve [90], and it is out of the scope of this very rudimentary analysis.

3.4.2   Coherent Integration

This section quantifies the loss of coherent SNR due to a constant Tx-Rx frequency offset. The derivation for coherent integration gain is first reproduced from Richards [90]. Then, the loss of coherent SNR due to a frequency offset is derived. From [90] a signal can be expressed as

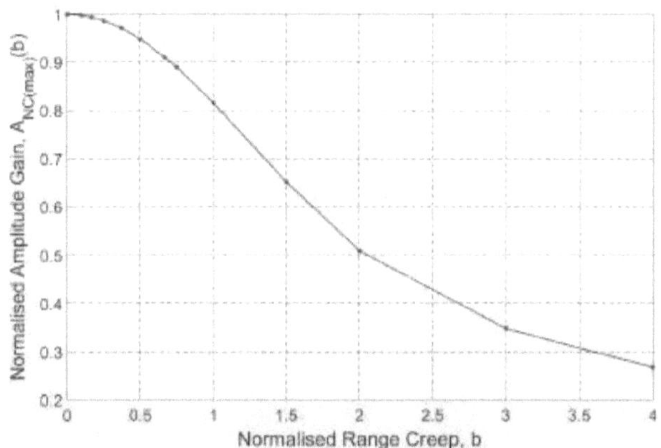

Figure 3.10: Non-Coherent Amplitude Gain vs. Range Creep. A range creep of a 100% of the 3 dB main lobe width decreases the non-coherent amplitude gain by about 20%.

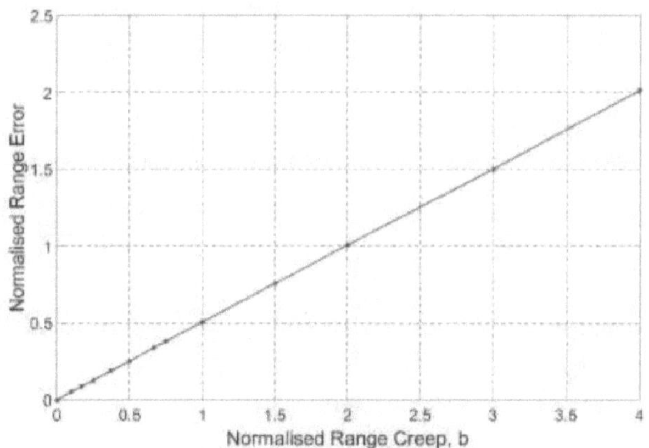

Figure 3.11: Non-Coherent Range Error vs. Range Creep. A range creep of a 100% of the 3 dB main lobe width results in a range error of 50%.

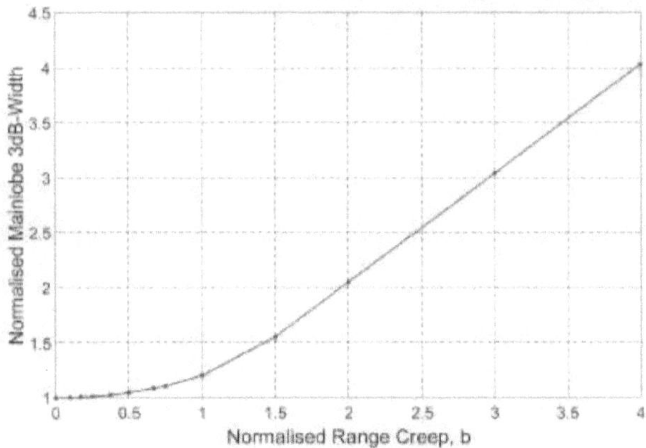

Figure 3.12: Non-Coherent 3-dB Width of Target Amplitude vs. Range Creep. A range creep of a 100% of the 3 dB main lobe width increase the perceived main lobe width by about 25%.

$$x[n] \quad = \quad s[n] + w[n] \tag{3.50}$$

$$= \quad A\sin\theta + w[n], \tag{3.51}$$

where s[n] has a constant amplitude of A for $\theta = \pi/2$ and is independent of n, and w[n] is white Gaussian noise with variance of σ_w^2. The SNR of a single sample is given by $\chi_1 = A^2/\sigma_w^2$. From [90] the coherent integrated signal can be written as

$$z \quad = \quad \sum_{n=0}^{N-1} \left(s[n] + w[n] \right), \tag{3.52}$$

$$\tag{3.53}$$

where the post integration signal and noise power is given by $P_s = (NA)^2$ and $P_w = N\sigma_w^2$, respectively. Thus, the post integration SNR is given by $\chi_{N_c} = NA^2/\sigma_w^2$. Thus,

$$\chi_{N_c} = N\chi_1, \tag{3.54}$$

and ideal coherent integration results in a SNR improvement of factor N, where the signal phase is constant.

However, in the case where there is a constant Tx-Rx frequency offset, the phase of the recorded signal advances linearly from pulse to pulse such that

$$s_\theta[n] = A\sin\left(\frac{\pi}{2} + \frac{n\theta}{2}\right), \tag{3.55}$$

and

$$z_\theta = \sum_{n=0}^{N-1}\left[A\sin\left(\frac{\pi}{2} + \frac{n\theta}{2}\right) + w[n]\right]. \tag{3.56}$$

The noise power in both cases remains identical, thus the reduction in coherent SNR due to a Tx-Rx frequency offset can be written as

$$\text{SNR}_{loss} = 1 - \frac{P_\theta}{P_s}. \tag{3.57}$$

This coherent SNR loss due to phase creep is plotted in Figure 3.13. A total phase creep of 95°, 160° and 266° results in coherent SNR loss of 1 dB, 3 dB and 10 dB, respectively. After 266° the SNR loss deteriorates rapidly. The loss tends to infinity at multiples of 360°.

Figure 3.14 shows the phase creep versus carrier frequency during one second of integration for various frequency offsets. Horizontal lines indicate SNR losses of 1 dB, 3 dB and 10 dB, respectively. If a 1 dB SNR loss at 2.4 GHz can be tolerated, then a frequency offset of $y_o \leq 10^{-10}$ is required. However, at 22 GHz the requirement becomes more stringent such that $y_o \leq 10^{-11}$. If only a few degrees of phase creep can be tolerated at 2.4 GHz then the Tx-Rx pair must be synchronised to $y_o \leq 10^{-12}$.

Thus, the goal is achieved, and one can now directly relate coherent SNR loss to the bistatic frequency offset.

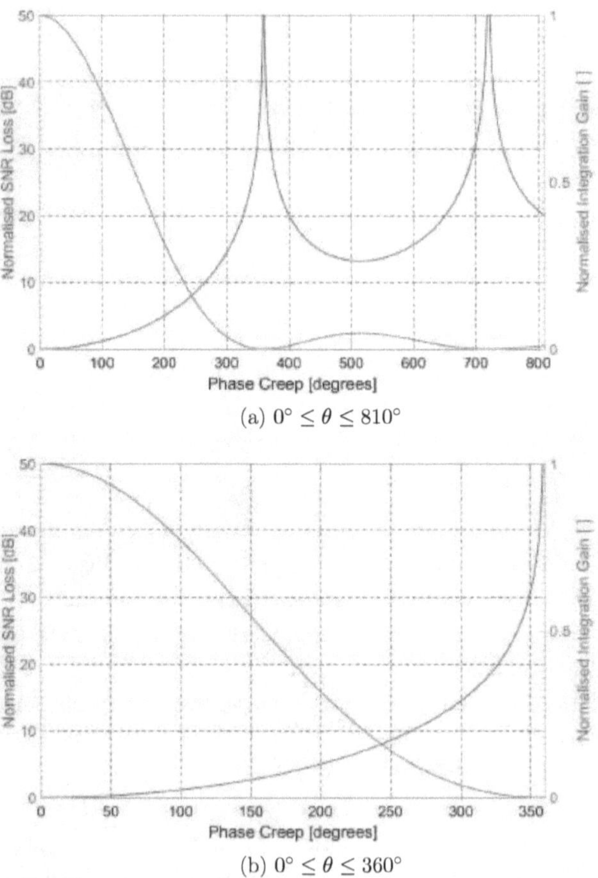

(a) $0° \leq \theta \leq 810°$

(b) $0° \leq \theta \leq 360°$

Figure 3.13: Coherent SNR Loss versus Total Phase Creep. A total phase creep of 95°, 160° and 266° results in coherent SNR loss of 1 dB, 3 dB and 10 dB, respectively. After 266° the SNR loss deteriorates rapidly. The loss tends to infinity at multiples of 360°.

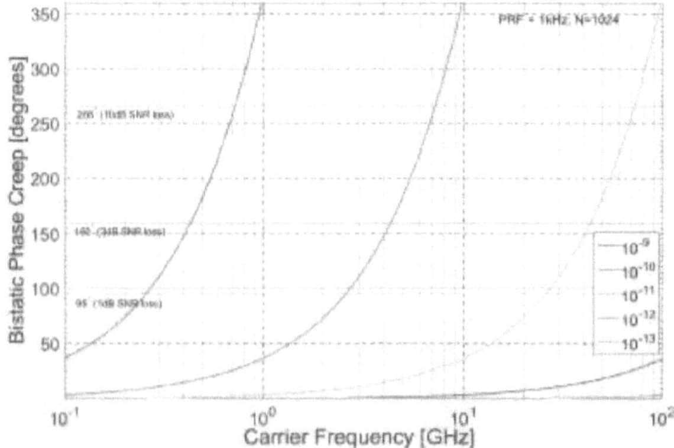

Figure 3.14: Phase Creep versus Carrier Frequency for Various Frequency Offsets. ($K_T = 0$, $D = 0$, $\sigma_y = 0$) Horizontal lines indicate SNR losses of 1 dB, 3 dB and 10 dB, respectively. If a 1 dB SNR loss can be tolerated, then a frequency offset of $y_o \leq 10^{-10}$ is required at 2.4 GHz. However, at 22 GHz the requirement becomes more stringent such that $y_o \leq 10^{-11}$. If only a few degrees of phase creep can be tolerated at 2.4 GHz then the Tx-Rx pair must be synchronised to $y_o \leq 10^{-12}$.

3.5 Bistatic Close-in Phase Noise

This section considers the phase noise requirements of bistatic systems. Section Section 2.2.4 reviewed the literature comparing bistatic to monostatic phase noise. Moreover, in monostatic systems, the LO phase noise is at least partially correlated during up- and down-conversion resulting in a partial phase noise cancellation at the IF. However, for bistatic systems using matched but independent STALOs, the Tx-Rx phase noise is entirely uncorrelated offering no IF phase noise cancellation. Thus, uncorrelated bistatic systems have a much more stringent phase noise requirement than monostatic systems if a similar SCV is required. From (2.6) and (2.7), one may express the difference between quasi-monostatic and monostatic phase noise such that

$$\Delta S_{n_{MB}} = S_{n_M} - S_{n_B} \tag{3.58}$$

$$= 2M^2[2\sin(\pi f_m \tau_{delay})]^2 S_{\phi_M}(f_m) - 4M^2 S_{\phi_B}(f_m), \tag{3.59}$$

where S_{n_M} is the monostatic IF phase noise, S_{n_B} is the bistatic IF phase noise, M is the ratio of the STALO to the carrier frequency, $S_\phi(f)$ is the single sideband STALO phase noise, f_m is the frequency offset from the carrier, and τ_{delay} is the monostatic propagation delay. Note that the difference, $\Delta S_{n_{MB}}$, is greater at shorter target ranges. Moreover, the difference $\Delta S_{n_{MB}}$ is relaxed by $20\,\mathrm{dBc/Hz}$ for every factor of ten increase in the target range. For an infinite target range $\Delta S_{n_{MB}} = 0$.

There are cases where the bistatic LOs are not entirely independent. In such cases, the LOs may be phase-locked to one another with a particular PLL bandwidth. Examples include LO synchronisation via GPS, an RF link, or fibre. These alternative methods were briefly reviewed in Section 2.3. Further, from Wolaver [79], it can be said that two phase-locked oscillators are well correlated for frequencies below the PLL bandwidth. Hence, it hints that one can approximate phase-locked Tx-Rx phase noise by replacing τ_{delay} in (2.6) with

$$\tau'_{delay} = \tau_B + \tau_{PLL}, \qquad\qquad [\mathrm{s}] \quad (3.60)$$

where τ_B is the total bistatic time delay, and τ_{PLL} is the PLL time constant.[1]

This intuitively implies that high bandwidth Tx-Rx PLL synchronisation will minimise the time delay, τ'_{delay}, and maximise phase noise cancellation. GPSDOs on the other hand, use very low PLL bandwidths where $\tau_{PLL} \gg 1$. Thus, it is likely that GPSDO synchronisation is best modelled with an infinite time delay. It is of interest to verify the validity of this approximation experimentally, but this is left to future work. However, the Doppler performance of the monostatic, bistatic, and LOS phase compensated bistatic cases are compared in Chapter 6 quantifying the achieved SCV for each case.

3.6   NetRAD Synchronisation Requirement

This section expands on the literature [3, 7–10, 20, 32] attempting to better specify the synchronisation requirements of bistatic pulsed-Doppler radar. Moreover, the error expressions derived in Section 3.2 are used to define a set of synchronisation targets for NetRAD. First, the relevant radar parameters are listed. Then, the STALO time, frequency and phase synchronisation targets are set based on

---

[1]Note that this is only a rough approximation assuming an abrupt transition between the reference and LO phase noises at the PLL output.

the desired radar range, Doppler and phase performance. The end goal is to determine if low-cost quartz-based GPSDOs are suitable to synchronise NetRAD. The typical expected hold-over performance of quartz and Rubidium standards are also compared.

3.6.1 Radar Parameters

This section briefly lists the NetRAD parameters required to derive its synchronisation requirements. Refer to Table 3.1.

NetRAD is a coherent tri-node pulsed-Doppler network radar. It has a carrier frequency of $v_{RF} = 2.4\,\text{GHz}$ and a bandwidth of $B = 50\,\text{MHz}$. The radar transmits a linear up-chirp waveform which is matched filtered at the receiver. The IF is at baseband such that $v_{RF} \approx v_{LO}$ where $v_{IF} \approx 0$ and each range bin matches the radar range resolution comprising 10 ns or 3 metres. In this text, the PRF is limited to 1 kHz. At this PRF, and at radar pulse lengths of $20.48\,\mu s$, the radar memory limits the maximum measurement time (target observation) to 130 seconds. The typical maximum coherent integration time is one second or less. Refer to Section 6.1 for more on the wireless NetRAD system.

3.6.2 Time

This section discusses the time synchronisation requirements of NetRAD. A typical bistatic time synchronisation requirement is one-tenth of the compressed pulse [3, 20]. For NetRAD, this is $\Delta\tau_B \leq 1/(10B) \leq 2\,\text{ns}$ or a range error of $\Delta R_B \leq 0.6\,\text{m}$. Further, the time induced bistatic range error, Δ_{R_B}, can be written from (3.35) and (3.37) as

$$\Delta_{R_B} = |-c\Delta\tau_B|, \qquad\qquad \text{[m]} \quad (3.61)$$

where, $\Delta\tau_B$, is the total cumulative Tx-Rx time error[1]. The STALO hold-over requirement is often specified by frequency stability, $\sigma_y(\tau)$, alone [3, 20]. However, this unrealistically assumes perfect time and frequency synchronisation whereas in practice the time error is often dominated by the frequency offset and drift. Also, the initial time offset is dependent on the time synchronisation mechanism (e.g. when using the GPS PPS signal or the direct breakthrough) which may be independent

---

[1]This is the relative Tx-Rx time error and may be negative.

Table 3.1: Basic NetRAD Specifications [27]

Parameter	Value
Carrier Frequency	2.4 GHz (S-band)
Bandwidth	50 MHz
Transmit Power	500 W (200 mW)
Antenna Gain	23 dBi
Antenna Beamwidth	10°
PRF[a]	50 Hz - 3 kHz
Pulse Length	0.1 µs - 10 µs
Waveforms[b]	Digitally Generated: Linear FM Chirp, Barker Codes, Polyphase codes, etc
Reference Frequency	100 MHz
Sampling Frequency	100 MHz
Nominal Range Resolution	3 metres
Range bin	3 metres (10 ns)

[a]The PRF is limited to 1 kHz throughout this text. This limits the maximum recording time to 130 seconds.

[b]Waveform is limited to a linear up-chirp throughout this text.

of the oscillator's phase (time). Hence, from (3.9) the bistatic STALO requirement may be better described by

$$|\Delta\tau_B| \geq |x_o| + |x_{drift}(\tau_h)| \tag{3.62}$$

$$\geq |x_o| + |y_o\tau_h + \Delta T K_T \tau_h + \frac{1}{2}DT_h^2 + \sigma_x(\tau_h)|, \qquad [\text{s}] \tag{3.63}$$

where, x_o, is the initial time synchronisation error, and $x_{drift}(\tau_h)$ is the total cumulative Tx-Rx time drift during the hold-over period, τ_h. The combined contributions of the time error, x_o, frequency offset, y_o, change in temperature, $\Delta T K_T$, frequency drift, D, and time instability, σ_x, must be less than or equal to $|\Delta\tau_B|$ during the hold-over period, τ_h. The contribution of each of the deterministic terms can be read from the plots in Figures 3.6 to 3.6.

From Figure 3.3, an OCXO synchronised to 10^{-12} is expected to drift by tens of nanoseconds per hour.[1] Conversely, a similarly synchronised Rubidium standard may drift less than a nanosecond in an hour. Hence, Quartz is possibly suitable to synchronise NetRAD for a few minutes of timing hold-over whereas a Rubidium standard may last for a couple of hours where $x_o = 0$.

Multi-channel common-view and carrier phase GPS time transfer techniques are reported to achieve the sub-2 ns time synchronisation required by NetRAD [19]. However, Figure 2.5 shows that two low-cost GPS receivers based on one-way time transfer can only maintain time synchronisation to within ±15 ns over multiple days across a baseline of 21.5 km. Nonetheless, this timing error reduced to less than ±5 ns after applying signal smoothing. This suggests that two low-cost GPSDOs should be able to maintain a relative timing accuracy of ±5 ns across short baselines. However, this assumes precisely surveyed antennas and no antenna temperature gradient. In practice, and for self-surveyed antenna positions, this timing accuracy is expected to be worse by a few nanoseconds.

GPSDO phase performance is often not published. Commercial GPSDOs may also behave unpredictably [31]. Consequently, it is not clear how the GPSDO phase steering may distort the short-term bistatic time difference during a 130-second target observation. Hence, the phase performance of individual GPSDOs requires calibration to verify adequate performance. However, it appears to be a reasonable assumption that a quartz GPSDO may wander by up to a few nanoseconds over a matter of minutes.

---

[1] One degree at 2.4 GHz equates to 1.16 ps.

Thus, a low-cost quartz GPSDO should achieve $x_o \leq \pm 5\,\text{ns}$ under ideal circumstances, and it is expected to drift by a further few nanoseconds during a 130-second recording. Hence, the initial range error will be greater than the NetRAD range resolution of 3 metres. However, it is expected to drift by less than the 10 ns range bin during a measurement. Therefore, it is worthwhile to investigate sidelobe breakthrough as a low-cost means to augment the time synchronisation capability of low-cost GPSDOs over short baselines.

3.6.3 Frequency

This section discusses the frequency synchronisation requirements of NetRAD. The target Doppler error due to imperfect frequency synchronisation is given by the integral of the Tx-Rx frequency offset over the coherent integration time. Hence, the initial frequency offset induces a Doppler error, and a frequency drift causes Doppler spreading during integration. [9] The Tx-Rx frequency synchronisation requirement may often be signal processing dependent. However, this text specifies NetRAD's required Tx-Rx frequency synchronisation as a function of the desired radial target velocity accuracy, $\mathbf{V_{cc}}$, and its Doppler resolution, y_{res}. The required Doppler accuracy, y_{acc}, is then given by

$$y_{acc} = \left| \frac{\mathbf{V_{acc}}}{c} \right| , \qquad [\ ] \qquad (3.64)$$

where c, the speed of light. Thus, the required initial Tx-Rx frequency synchronisation, y_o, is

$$y_o \leq y_{acc}. \qquad [\ ] \qquad (3.65)$$

NetRAD requires $\mathbf{V_{acc}} \leq 1\,\text{km/h}$ and consequently $y_o \leq 9.27 \times 10^{-10}$.
Further, the fractional Doppler resolution, y_{res}, is

$$y_{res} = \frac{\text{PRF}}{2Nv_{RF}} \qquad [\ ] \qquad (3.66)$$

where, v_{RF}, is the carrier frequency and, N, is the number of pulses (fast Fourier transform samples). It is required that the Tx-Rx frequency drift less than y_{res} over the integration period, τ_{int}, such that

$$y_{res} \geq |\Delta y_{drift}(\tau_{int})| \qquad (3.67)$$

$$\geq |\Delta T(\tau_{int})K_T + D\tau_{int} + \sigma_y(\tau_{int})|, \qquad |\ | \quad (3.68)$$

where, $\Delta y_{drift}(\tau_{int})$, is the total cumulative Tx-Rx frequency drift. The combined contributions of temperature drift, $\Delta T K_T$, frequency drift, D, and frequency instability, σ_y, must be less than or equal to the Doppler resolution. Refer to Figures 3.4 and 3.5.

Most GPSDO applications are concerned with the long-term performance where the frequency is averaged for extended periods. However, for Doppler radar, the short-term GPSDO frequency accuracy is most important. Moreover, it is the average Tx-Rx frequency offset, as well as, the frequency drift, about the mean, at the time of, and during, the target measurement, that is of concern. However, the short-term frequency accuracy is not often specified and must be user calibrated.

A one-second fast Fourier transform (FFT) where $N = 1000$ and PRF $= 1\,\mathrm{kHz}$ produces a Doppler resolution of $y_{res} = 2.1 \times 10^{-10}$. Hence, a one-second averaged GPSDO frequency plot should have peak-to-peak accuracy of better than 2.1×10^{-10} to synchronise NetRAD adequately. There are published results showing a short-term accuracy of a few parts in 10^{-11} for a low-cost GPSDO [75] which exceeds the NetRAD requirement by about an order of magnitude. Also, during hold-over, typical OCXO and Rubidium standards may age by a few parts in 10^{-10} and 10^{-12} per day, respectively. Hence, quartz should be able to maintain sufficient frequency hold-over for a few hours to a day whereas Rubidium may last for a week or more. However, the short-term frequency noise may limit the performance.

3.6.4 Phase

This section discusses the phase synchronisation requirements of NetRAD. The phase requirements of monostatic and bistatic systems are similar and may range from less than a degree to tens of degrees depending on the processing technique and duration [3]. However, Hurley et al.[36] has also shown that a coherent network radar system relying on constructive interference at the target may have additional carrier phase requirements. In this text, the tri-node NetRAD is modelled as two independent bistatic channels with no interference at the target.

Absolute Tx-Rx phase is often not essential and is dependent on the bistatic target range. However, one may define the maximum Tx-Rx time drift during integration, $\Delta\tau_\Phi$, in terms of the maximum allowable phase drift, $\Delta\Phi_{max}$, as

$$\Delta\tau_\Phi = \frac{\Delta\Phi_{max}}{2\pi v_{RF}}, \qquad\qquad \text{[s]} \quad (3.69)$$

where, v_{RF}, is the carrier frequency. Hence, it is required that

$$
\begin{aligned}
\Delta\tau_\Phi \;\geq\;& x_{drift}(\tau_{int}) & (3.70)\\
\geq\;& \left| y_o\tau_{int} + \Delta_T K_T \tau_{int} + \frac{1}{2}D\tau_{int}^2 + \sigma_x(\tau_{int}) \right|, & \text{[s]} \quad (3.71)
\end{aligned}
$$

where the combined contributions of the frequency offset, y_o, change in temperature, $\Delta T K_T$, frequency drift, D, and time instability, σ_x, must be less than or equal to the allowed time drift over the integration period, τ_{int}. Figure 3.3 compares the hold-over phase performance of hypothetical frequency standards at 2.4 GHz. Moreover, quartz, Rubidium and Cesium standards synchronised to 10^{-12} may drift by 400°, 4° and 1°, respectively, during one second.

The STALO hold-over requirement is often specified by frequency stability, $\sigma_y(\tau)$, alone [3, 8]. However, in practice, and for high-performance STALOs, the time drift may be entirely dominated by a constant frequency offset for short integration periods of a few seconds or less. Thus, one may instead specify the maximum integration frequency offset, Δy_Φ, such that

$$
\begin{aligned}
\Delta y_\Phi \;=\;& \frac{\Delta\tau_\Phi}{\tau_{int}} & (3.72)\\
\geq\;& |y_o + \Delta_T K_T|. & [\,] \quad (3.73)
\end{aligned}
$$

where $|D\tau_{int} + \sigma_x(\tau_{int})| \ll |y_o + \Delta_T K_T|$.

In this text, NetRAD is limited to the simple, coherent pulse integration described in Section 3.4.2. Thus, a maximum integration phase drift of 95° (110 ps), 160° (185 ps) and 266° (266 ps) is expected to result in an SNR loss of 1 dB, 3 dB and 10 dB, respectively. Then, from Figure 3.14, it is required that $\Delta y_\Phi \leq 10^{-10}$ for a maximum coherent SNR loss of 1 dB during one second of integration.

In the case of GPSDO synchronisation, the frequency and temperature offsets

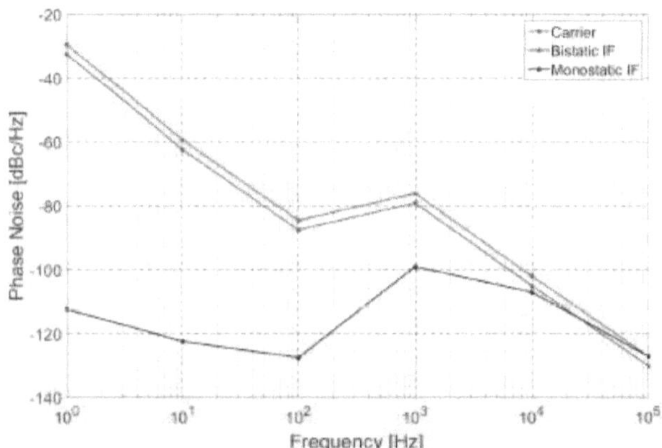

Figure 3.15: Predicted Close-In IF Phase Noise for NetRAD: Bistatic versus Monostatic. ($\tau_{delay} = 20\,\mu\text{s}$.) The monostatic cancellation of $20\,\text{dBc/Hz}$ per decade transforms the Flicker FM noise at the carrier to Flicker PM noise at the IF. The bistatic IF phase noise increase by $3\,\text{dBc/Hz}$.

and drifts are constantly nulled through active frequency steering. A PLL is a low-pass filter of phase [79], and it is expected that the rate of change of phase will decrease with increasing PLL time constants. If it is assumed that the phase slope is constant during the bistatic pulse integration, then the instantaneous Tx-Rx offset at the time of recording must be less than Δ_{y_Φ}. Thus, the short-term GPSDO frequency accuracy must be less than Δ_{y_Φ}. Low-cost GPSDOs have been reported to achieve a short-term accuracy of a few parts in 10^{-11} which is adequate to limit the NetRAD coherent SNR loss to $1\,\text{dB}$ for one second of integration.

3.6.5 Close-In Phase Noise

This section discusses the close-in phase noise requirements of NetRAD. In Section 2.2.4 and Section 3.5, it was discussed how the monostatic phase noise cancels to a large extent at the IF. Conversely, for GPSDO synchronisation, the bistatic phase noise adds and increase by $3\,\text{dBc/Hz}$ at the IF.

Ultra-low close-in phase noise STALOs are very expensive. Thus, it was decided to set the carrier phase noise requirement for the wireless GPSDO synchronised

NetRAD similar to that of the cabled NetRAD system.

NetRAD PLL synthesises its 2.4 GHz carrier directly from its 100 MHz STALO. This PLL has a bandwidth of 2 kHz [12], and the STALO entirely contributes the phase noise within this bandwidth [79]. Hence, lowering the STALO close-in phase noise appears to be the only way to improve the bistatic phase noise for GPSDO synchronised systems autonomously. There was never the opportunity to measure the carrier phase noise of the GPSDO synchronised NetRAD. However, one can make a fair estimation of the carrier phase noise using the 100 MHz UCT GPSDO phase noise measured in Section 5.5 and Derham's [12] account of the NetRAD 2.4 GHz PLL filter.

Figure 3.15 predicts the IF phase noise of NetRAD when synchronised with the UCT GPSDOs using the above-estimated carrier phase noise. The predicted monostatic and bistatic phase noises at the IF are plotted for $\tau_{delay} = 20\,\mu s^{1}$ from (2.6) and (2.7). The monostatic noise starts to cancel at 20 dBc/Hz per decade at frequencies below 25 kHz. This transforms the Flicker FM noise at the carrier to Flicker phase modulation (PM) noise at the IF. The bistatic IF phase noise increase by 3 dBc/Hz for all frequencies. There is a severe discrepancy between the monostatic and bistatic phase noises. At 1 Hz the difference exceeds 80 dBc/Hz. Consequently, there is a severe penalty in bistatic SCV.

In Chapter 6, LOS phase compensation is used in an attempt to suppress the increased bistatic phase noise to improve the bistatic SCV at low cost.

3.7 Conclusion

This chapter derived the effect of imperfect bistatic Tx-Rx synchronisation on pulsed-Doppler radar performance from basic principles. Moreover, the sampled IF signal was expressed as a function of Tx-Rx synchronisation, radar pulse number, ADC sample number, and an arbitrary future time. It built on the existing literature [3, 7–10, 20, 32] and related STALO synchronisation directly to the bistatic radar performance, and provided a set of plots allowing a designer to determine the required level of synchronisation for a specific application rapidly.

It identified that the Tx-Rx phase drift is often dominated by a constant frequency offset for high-performance STALOs and integration periods of a few seconds

---

[1]A time delay of 20 ns equates to a monostatic range of 3 km.

or less. Bistatic pulse integration in the presence of such a constant frequency offset was studied. Moreover, it was shown that a frequency offset decreases the amplitude gain, broadens the main lobe, and introduces a range error during non-coherent integration. Further, during coherent integration, a total linear phase drift of 95°, 160° and 266° cause an SNR loss of 1 dB, 3 dB and 10 dB, respectively.

The bistatic timing requirement was defined as the allowable initial time offset, x_o, and the allowable time drift, $x_{drift}(\tau_h)$, during a hold-over period, τ_h. Similarly, the bistatic frequency requirement was specified as the allowable initial frequency offset, y_o, which is determined by the desired radial target velocity accuracy, $\mathbf{V_{cc}}$, and the allowable frequency drift, $\Delta y_{drift}(\tau_{int})$, during a Doppler integration period, τ_{int}, which is determined by the desired Doppler resolution, y_{res}. The literature [3, 8] often assumes perfect Tx-Rx synchronisation and considers STALO instability as the only source of error. However, perfect synchronisation is unrealistic, and synchronisation is better specified in terms of bistatic time drift which is a function of frequency offset, temperature offset, oscillator ageing, and time stability. It is further recognised that Tx-Rx frequency offset may often dominate for high-performance STALOs over short periods. In such cases, a Tx-Rx frequency offset may be specified to meet a desired bistatic phase requirement.

One-way GPS time transfer is inadequate to synchronise NetRAD to within a fraction of its 3 m range resolution. Low-cost GPSDOs are expected to achieve a short baseline range error of ± 1.5 m when assuming that all biases are nulled, the antennas are precisely surveyed, and that there are no antenna temperature gradients. The range accuracy is expected to be worse by a few metres under field conditions. However, low-cost one-way GPSDOs are expected to provide a short baseline short-term frequency accuracy of better than approximately 10^{-10}. Thus, NetRAD should easily achieve a $\mathbf{V_{cc}} \leq 1$ km/h with a Tx-Rx frequency drift less than the Doppler resolution for integration periods of one second. It should also maintain an SNR loss of below 1 dB for coherent bistatic integration lasting a few seconds. The bistatic SCV is estimated to be poor. Moreover, the STALO phase noise must be improved by approximately 80 dBc/Hz to achieve a bistatic SCV similar to that of a monostatic radar for a target at 3 km. However, it is postulated that LOS phase compensation may be an inexpensive method to improve both the time synchronisation and bistatic SCV across short baselines.

In Chapter 4, a set of three low-cost quartz GPSDOs based on one-way GPS time transfer are designed and built. They are then calibrated in Chapter 5 using the low-cost DMTD system designed in Appendix C. In Chapter 6, these GPSDOs are used to synchronise NetRAD, and then the real bistatic data is compared to what was predicted in this chapter. The effect of LOS phase compensation on phase, frequency, phase noise and Doppler performance is also quantified using the real data.

Chapter 4

Design of the UCT GPSDO

In this chapter, a set of three identical low-cost quartz GPSDOs are designed and built [34, 35]. These GPSDOs are calibrated in Chapter 5 using the DMTD system designed in Appendix C and then later used to synchronise the three NetRAD nodes in Chapter 6. They were explicitly developed as research tools to investigate network radar synchronisation. They solved three practical difficulties. Firstly, they are capable of network-wide time synchronisation that is up to an order of magnitude more accurate than what is possible with the raw GPS PPS signals. Secondly, a quick-locking adaptive PLL is implemented with a lock-in time up to four times quicker than a conventional PLL. Thirdly, the GUI includes a development environment where sophisticated control loops can be rapidly developed on a PC for verification directly on the GPSDO hardware.

The UCT GPSDO can be described as a hybrid PLL where everything except the STALO is implemented digitally. Refer to Figure 4.1. Here, a low-cost Motorola M12+ timing receiver generates a PPS time reference. The 10 MHz output of a single-oven Oscilloqaurtz 8788 OCXO is digitally divided to 1 MHz. The 8788 is an ultra-low noise OCXO with a reasonable ageing rate of 5×10^{-10}/day. Then, an ACAM TDC-GP2 high-resolution TDC measures the time difference between each PPS rising edge and the closest corresponding 1 MHz rising edge. The TDC-GP2 is a 65 ps resolution delay line based TDC offering an LSB frequency resolution of 6.5×10^{-11}. The 16-bit phase error drives an FPGA-based PLL filter. The computed filter output sets the DAC voltage which steers the OCXO to complete the phase-locked loop. The 10 MHz output is analogue multiplied by 10 to 100 MHz and distributed in LVPECL and LVDS digital formats. The 32-bit PLL filter module is implemented in a Xilinx Spartan3 FPGA. It consists of five stages. Each step is optional, with the user able to enable or bypass any particular section. The first stage is the real-time sawtooth correction where the 8-bit sawtooth error, supplied by the GPS receiver, is subtracted from the measured phase error. Outliers are removed in the second step to prevent glitches from erroneously biasing the PLL filter memory. However, the firmware outlier removal is, to date, not fully implemented. The third section includes a moving average prefilter. Often the high-jitter phase error is of zero mean, and a prefilter may be beneficial in some instance. The fourth section contains an infinite impulse response (IIR) PLL filter with user-specifiable coefficients. Finally, the anti-windup limiter prevents the PLL filter from integrating past the DAC limits.

The firmware is fully autonomous allowing the GPSDO to operate without external intervention. It is implemented in hardware description language (HDL) and consists of a custom picoBlaze based parallel multi-microcontroller architecture. This parallel architecture relaxes the sequential timing requirements and makes the HDL modular and easily extensible. The designer is free to choose between parallel HDL or sequential assembler code which reduces the design time. New modules can be added on an ad-hoc basis without influencing the existing HDL design. The GPSDO is reconfigurable via a simple serial interface allowing real-time access to the low-level GPSDO parameters. Moreover, the PLL filter coefficients are arbitrarily configurable. The EP firmware module solves the problem of network-wide time synchronisation. Moreover, the EP module produces an accurate GPS synchronised pulse at an arbitrary user-selected future time. The EP offers the best estimate of true UTC at any given moment, and it is up to an order of magnitude more accurate than the raw GPS PPS output.

The GPSDO GUI enables real-time monitoring and control of up to three GPSDOs via their respective RS-232. Diagnostic data such as PLL phase errors and DAC steering voltages can be plot and recorded in real-time. This GUI includes a smart PLL filter module. This module serves as a development environment where sophisticated PLL filters can be rapidly developed on a PC for verification directly on the GPSDO hardware. This environment was used to develop and test outlier removal and phase-locked detection algorithms. It also assisted in the development of a quick-locking adaptive PLL able to achieve phase lock up to four times faster than a conventional PLL.

The following sections discuss the GPSDO hardware, firmware and software.

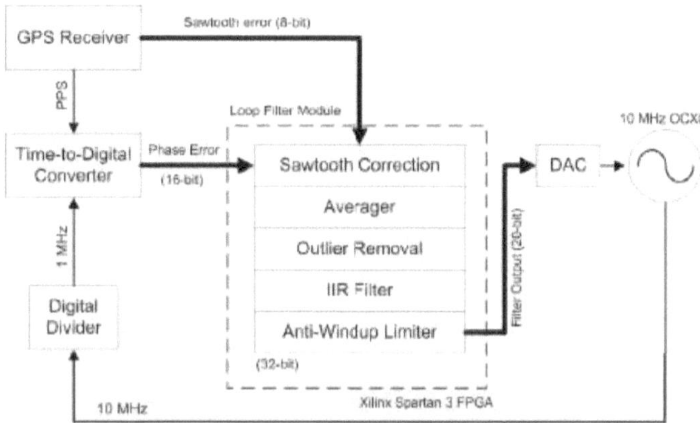

Figure 4.1: UCT GPSDO PLL Architecture.

4.1 Hardware

This section describes the hardware and subsystems of the UCT GPSDO. The block diagrams in Figures 4.1 and 4.2 provide a system overview. Moreover, a Motorola M12+ GPS receiver generates a PPS edge that is traceable to UTC. Simultaneously, the 10 MHz output of an Oscilloquartz 8788 OCXO is digitally divided down to 1 MHz. Now, the time difference between the zero-crossings of this 1 MHz and that of the GPS PPS is measured using an ACAM TDC-GP2 TDC. This 16-bit phase error drives an FPGA based PLL filter. The computed filter output then sets a 20-bit DAC which steers the OCXO to complete the phase-locked loop. Central to the UCT GPSDO architecture is a Xilinx Spartan 3 FPGA. This FPGA manages the various subsystems, including the GPS receiver, phase detector, the DAC frequency steering, user communication, the clock distribution, and the power supply unit. The UCT GPSDO is based on the PC-104 form factor and consists of a stack of four PCBs. See Figure 4.3 for photos of the front and rear panel, as well as the different PCB modules. The remainder of this section covers the various hardware components and circuitry in more detail.

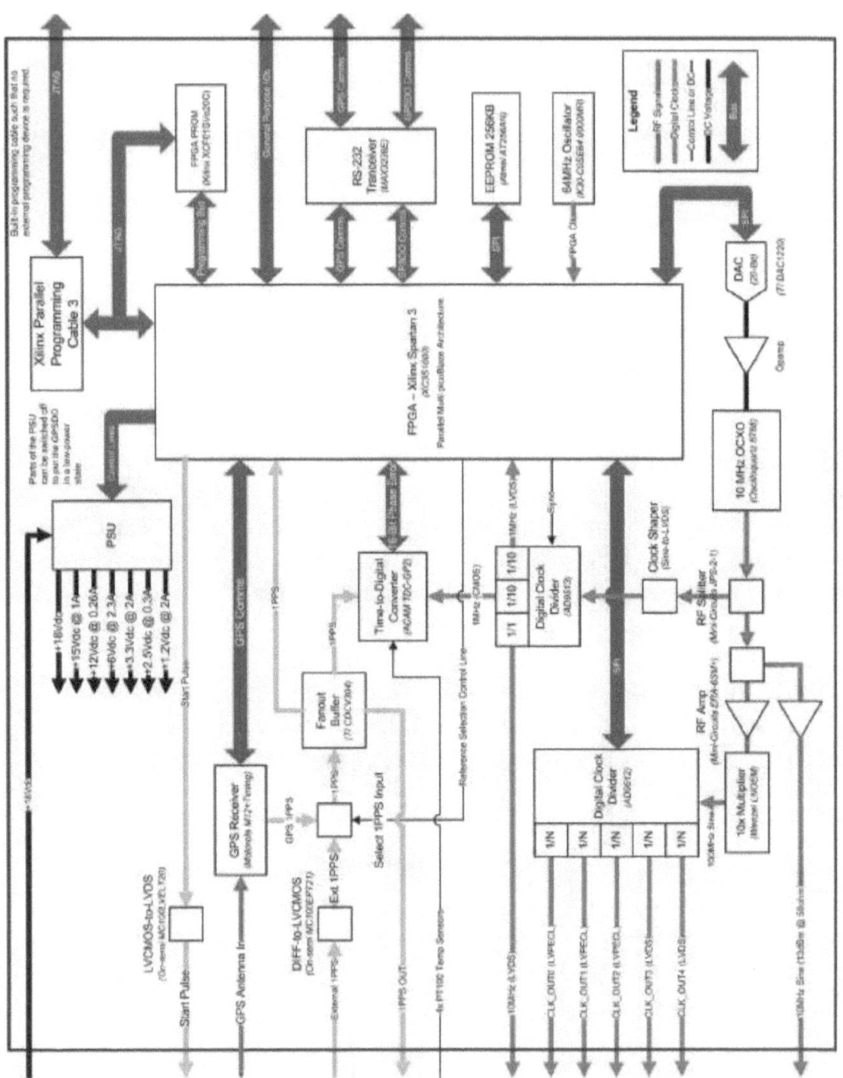

Figure 4.2: Systemic overview of the UCT GPSDO.

(a) Enclosure Front View

(b) Enclosure Rear View

(c) Power Supply Unit

(d) M12+ and Digital Circuitry

(e) OCXO and Frequency Steering

(f) Multiplier and Clock Distribution

Figure 4.3: The UCT GPSDO

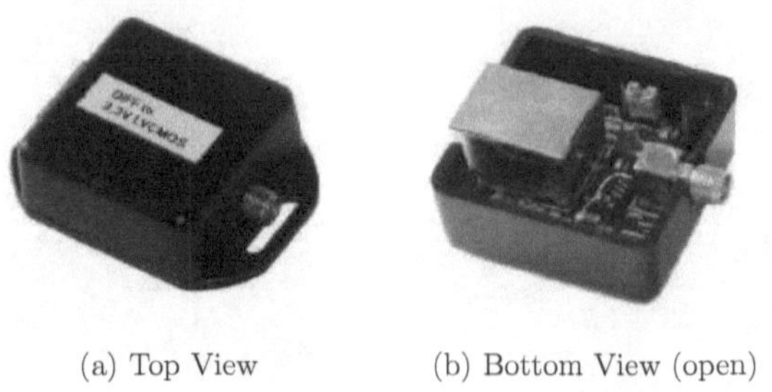

(a) Top View   (b) Bottom View (open)

Figure 4.4: Differential to Single-End Logic Level Translator

4.1.1   GPS Antenna and Receiver

The Motorola M12+ timing GPS receiver and the Motorola Timing2000 GPS antenna were chosen for the UCT GPSDO. Both are low-cost specialist timekeeping devices that are commercially available. Hambly and Clark [60] calibrated four M12+ receivers against the United States Naval Observatory (USNO) Master Clock and published detailed performance results. This data made the M12+ the preferred choice over other GPS timing receivers that were available at the time. Refer to Section 2.4.4 and Section 2.4.2 where low-cost timing receivers and their performances are discussed. The M12+ is visible in Figure 4.3d as the credit card sized mezzanine board. The PPS sawtooth correction is applied in real-time by the GPSDO firmware (see Section 4.2.3).

4.1.2   Phase Detection

The high-resolution ACAM TDC-GP2 TDC was chosen for phase detection. The TDC-GP2 is a delay line based TDC which measures propagation delay through calibrated on-chip logic buffers. It has a 65 ps LSB resolution and RMS noise of 50 ps. Hence, the TDC-GP2 offers an LSB frequency resolution of 6.5×10^{-11} for a once per second update rate. A time resolution below 1 ns is likely adequate for single carrier GPS time transfer. However, the 65 ps resolution is convenient when phase-locking to a low jitter external PPS reference. Often the STALO frequency is divided to a much lower frequency for phase detection. In some applications, it is divided by 10^7 all the way down to 1 Hz. However, the process of digital division is similar to sampling, and the noise and spurs fold back into the spectrum at high division ratios. This folding increases the output jitter significantly [91]. The TDC-GP2 offers the advantage of direct measurement at 10 MHz with no digital division required. In this application, the PLL-steered 10 MHz OCXO is divided by 10 to 1 MHz to increase the detection range to 1 μs. The TDC-GP2 measures the time difference between this 1 MHz signal and the PPS edge. The high 65 ps resolution came at the cost of a limited 1.8 ms measurement range. However, a larger measurement range is preferred. The phase detection circuitry is located on the digital PCB shown in Figure 4.3d. Refer to the block diagram in Figure 4.2.

The Analog Devices AD9513 does the clock division off-chip to the FPGA. The AD9513 has well-specified jitter, phase noise and propagation delays whereas FPGA jitter and propagation delays are HDL dependent. A clock shaper[1] low-noise amplifies the 1 MHz zero-crossing slopes providing a high slew rate clock input to the AD9513. The FPGA controls the TDC-GP2 via an serial peripheral interface (SPI) bus. Moreover, the FPGA gates the TDC-GP2 operation to keep the PPS edge at the midpoint of the 1.8 μs measurement range. The 64 MHz FPGA clock measures the previous PPS period to predict when then next PPS edge will occur. Thereby centring the PPS edge ensuring a ± 500 ns, or $\pm \pi$ rad, detection range. Also, the 1 MHz signal has a period of 1 ms which guarantees a single rising edge within the 1.8 ms measurement range. The 16-bit TDC result is read and processed by the FPGA.

The user has the option to select either the internal M12+ PPS or an external PPS input. The external PPS input in combination with the GPSDO PPS output can be used to phase-lock the three UCT GPSDOs to each other. The result is a free-standing and high-performance copper synchronisation solution. The external PPS input and high-resolution phase detection were utilised to test the CERN WR fibre synchronisation system [41, 42]. Further, both the PPS and the 1 MHz clock signals are routed to the FPGA. This leaves the option to study firmware FPGA-based phase detection using the more conventional XOR or Flip-Flop methods.

4.1.3 STALO and Steering Circuitry

A GPSDO is self-calibrating where active GPS-steering nulls the combined offsets and long-term drifts within the GPSDO circuitry [31]. Nonetheless, the STALO and steering circuitry must provide hold-over for at least as long as the PLL time

---

[1]Circuitry was found on Dr Griffiths' website [92].

constant and preferably longer. The longer the PLL time constant, the greater the demand is on the circuit stability. Low-cost quartz GPSDOs have typical time constants ranging from a few hundred to a few thousand seconds. Therefore, most of the UCT GPSDO is implemented digitally to guard against environmental effects. However, the frequency steering remains mostly analogue and is sensitive to the ambient. Careful design is required to minimise noise and long-term drift within the STALO steering. [73] This section discusses the UCT GPSDO's STALO and its steering circuitry.

The 10 MHz Oscilloquartz 8788 SC-cut single oven OCXO was chosen as the STALO. The 8788 is low-cost with moderate performance. It has an ultra-low phase noise of -110 dBc/Hz at 1 Hz and a -160 dBc/Hz noise floor. It is reasonably stable with $\sigma_y(\tau = 1) \leq 10^{-12}$ and ages less than 5×10^{-10}/day. The 8788 can be steered $\pm 8 \times 10^{-7}$ by applying 5 ± 5 V to its EFC pin. Consult Appendix B for the full specification of the Oscilloquartz 8788.

The GP2-TDC phase detector (Section 4.1.2) offers an LSB frequency resolution of 6.5×10^{-11}. The OCXO circuitry must be stable to at least within this LSB detection resolution for the duration of τ_{PLL}. However, the PLL filter further increases the frequency resolution by smoothing the TDC output. Hence, an even finer steering resolution is preferred. A 16-bit or 20-bit DAC paired to the 8788 offers LSB steering resolutions of 2.4×10^{-11} or 1.5×10^{-12}, respectively. There was never an opportunity to measure the performance of the steering circuitry. Further, due to project time constraints, the original prototype was used in the end. Consequently, the steering circuitry is overcomplicated and leave much room for improvement. Nonetheless, the UCT GPSDOs had adequate performance for the radar measurements in Chapter 6. However, it is unlikely that the steering is stable to the full 20-bit LSB resolution for $\tau_{PLL} = 1000$ s, but the author is reasonably confident that the circuit is stable at least to within the TDC LSB resolution and possibly even better.

The 20-bit TI DAC1220 was chosen for its low noise, high linearity and ultra-high resolution. It also has a built-in self-calibration function to further improve linearity. The DAC1220 is pulse width modulated (PWM) based and controlled via a two-wire SPI interface. The SPI lines are digitally isolated using the Analog Devices ADUM1251 at an attempt to prevent unwanted ground loops. A Maxxim MAX6325 buried Zener voltage reference references the DAC1220. The MAX6325 was chosen on its low noise and ultra-low temperature coefficient of 1 ppm/°C. However, it was later discovered that the 8788 provides an oven-stabilised voltage reference which would have been a far superior choice. A low-noise OPA27 opamp buffers the DAC1220 output and drives the 8788 EFC. This OPA27 is configured as an active low-pass filter with a 20 Hz cut-off to lower the EFC noise bandwidth. Various low noise linear regulators supply the 8788, DAC1220, MAX6325 and OPA27 which all have very good PSSRs. The 8788 drifts by $\pm 2 \times 10^{-10}$ for a supply variation of 12 V±5%. Hence, if one considers only the 8788's contribution, the supply variation must be less than 195 mV, 73 mV, or 4.6 mV to stay within the TDC, 16-bit, or 20-bit LSB resolution, respectively.[1] Similarly, the circuitry was designed to be temperature stable. The 8788 drifts by 4×10^{-10} over it 60°C operating range. Hence, if one considers only the 8788's contribution, the enclosure temperature must vary less than 9.8°C, 3.7°C or 0.23°C to stay within the TDC, 16-bit, or 20-bit LSB resolution, respectively.[2] The PCB containing the OCXO and frequency steering board is shown in Figure 4.3e.

Recommended future improvements include better buffering, isolation and filtering for the 8788 10 MHz output. The voltage standing wave ratio (VSWR) stabilising techniques discussed in Appendix C should be applied to the 10 MHz clock distribution. The lower noise LT1007 should replace the OPA27. Also, the filters limiting the noise bandwidth at the 8788 EFC input and MAX6325 output must be improved. Further, large ceramic capacitors with superior frequency response and lower leakage should replace the existing noisy, temperature sensitive, and high leakage tantalum and foil electrolytic capacitors. However, care must be taken with DC voltage biases across the ceramic dielectrics. The steering circuitry itself should also be enclosed within an electromagnetic interference (EMI) shield.

4.1.4 Clock Distribution

The clock distribution board, depicted in Figure 4.3f, serves as a clock multiplier and digital fanout buffer for the 10 MHz GPS-steered OCXO output. Each NetRAD node requires a 100 MHz LVPECL frequency reference. Hence, the 10 MHz OCXO is multiplied by ten to 100 MHz using a Wenzel LNOEM analogue multiplier. Moreover, an RF splitter splits the sinusoidal 10 MHz OCXO input 1:2. Both splitter outputs are individually buffered using a Mini-Circuits ERA-6SM+ MMIC RF

---

[1]The power supply variation is assumed linear across the voltage range.

[2]For simplicity, temperature drift is assumed to be linear. However, OCXO temperature drift is known to be non-linear.

amplifier. One 10 MHz output feeds the Wenzel multiplier, and the other is made available to the user. The AD9512 distributes the 100 MHz Wenzel output as three differential LVPECL signals and two differential LVDS signals. The AD9512 was chosen for its low and well-specified jitter and phase noise. Each clock output has an optional divider with an integer division of N where N = {1..32}. One LVDS output can be fine delay adjusted. The AD9512 is fully configurable via the GPSDO registers. Refer to Figure 4.2 for a block diagram.

The analogue multiplication came at the cost of increasing the phase noise by 20 dBc/Hz [93]. However, it achieves the goal of producing a phase noise level equal to or below that of the cabled NetRAD 100 MHz. Refer to Derham [12]. Digital PLL or DDS synthesis adds complexity and would have required more design time. A better and lower phase noise solution would have been to use a 100 MHz instead of a 10 MHz OCXO from the outset.

The ERA-6SM+ buffers were initially meant for testing and debugging only. Due to time constraints and imminent NetRAD trials, there was never time to populate the provisioned common-base bipolar junction transistor (BJT) amplifiers. These common-base buffers purposely have low phase noise and high reverse isolation.[1] Keeping the ERA-6SM+ buffers is regrettable because they have poor reverse isolation and raise the phase noise by 15 dBc/Hz to 20 dBc/Hz. Moreover, the GPSDO output phase noise ended up slightly higher than that of the cable NetRAD system when using the ERA-6SM+ buffers. Refer to Section 5.5. It is highly recommended to populate the common-base buffers for future experiments. Better RF filtering is also desirable. Refer to Tremblay and Tetu [94] to select an appropriate corner frequency for optimal frequency stability.

4.1.5   Power Supply

A custom designed power supply unit (PSU) powers the UCT GPSDO. The PSU consists of multiple high-efficiency DC/DC switchers which generate the various required voltage levels. The PSU accepts an $18V_{DC}$-$20V_{DC}$ input allowing it to be powered from mains using a standard AC/DC laptop brick. Readily available DC/DC laptop supplies enable it to run from an automotive $12V_{DC}$ source. The firmware can turn the individual voltages on and off to put the GPSDO in a low-power state. Care was taken to limit noise and EMI effects. The PCB is mounted

---

[1] Circuitry supplied by Dr Bruce Griffiths a 'time-nut'.

facing away from the other GPSDO circuitry to limit radio frequency interference (RFI). Some voltages are post regulated and filtered at the destination PCB to reduce noise and ripple further. See Figure 4.3c for a photo of the PSU.

4.1.6 Enclosure

This section details the UCT GPSDO enclosure. The front and rear panels are described and the choice of using the PC-104 form factor is discussed.

The front panel (Figure 4.3a) contains differential RJ45 clock connectors, status LEDs, a standby switch, a reset socket, and three DB9 connectors. The FPGA JTAG pins, RS232 communication, and some general IO lines are accessible via the three DB9 connectors. There are also optional differential/single-ended logic level translators which are powered directly from the front panel. See Figure 4.4. The standby switch puts the GPSDO into a low-power state leaving only the OCXO and other essentials running. During transportation, this mode extends the battery life while preserving the OCXO stability. The rear panel (Figure 4.3b) contains the DC power socket, fuse, on/off switch, and the SMA GPS antenna connector.

The GPSDO was initially meant to form part of an existing PC-104 based project. However, that project got discontinued and left the UCT GPSDO with the somewhat awkward form factor. The modular multi-PCB design is cumbersome with many board-to-board signals. It is difficult to probe signals with the stack assembled and impossible when inside the enclosure. A 19" form factor would have been better by reducing the number of internal connectors and eliminating much of the EMI control issues.

4.2 Firmware

The UCT GPSDO firmware is fully autonomous allowing the GPSDO to operate without intervention from external software. The firmware is implemented entirely within a Xilinx Spartan3 FPGA. It manages the onboard digital operations such as user communication, interfacing with the GPS receiver, phase detection, PLL filtering, frequency steering, and clock distribution. The firmware has a custom parallel multi-microcontroller architecture. Moreover, an arbiter manages and prioritises interrupt requests from and to multiple picoBlaze (pBlaze) processors communicating exclusively via a common set of system registers. This firmware is field upgradeable.

Moreover, the FPGA can be flashed from a PC's parallel port using the custom Xilinx programming Cable III built-in to the GPSDO.

The following sections first discuss the parallel firmware architecture. Then, the GPSDO RS-232 serial communication and the PLL filter module is described. Finally, the EP time synchronisation module is explained. This EP is GPS-synchronous but has the timing performance of the GPS-steered OCXO output. It enables independent UCT GPSDOs to establish an accurate network-wide time epoch.

4.2.1 Parallel Multi-Microcontroller Architecture

The parallel multi-microcontroller architecture enables multiple processors to work cooperatively as well as simultaneously. This system is interrupt based with an arbiter controlling and prioritising access to a common register map. Data transfer between processors happens exclusively via the system registers.

The pBlaze microcontroller was chosen because it is tiny, simple, free, and easy to learn. It is an 8-bit fully embedded microcontroller optimised for Spartan3 occupying only 96 slices and runs at 44 MIPS/88 MHz. It seamlessly integrates with VHSIC Hardware Description Language (VHDL) through its input and output ports. Its program memory can be updated in-situ without recompiling the VHDL code. The pBlazeIDE development environment and KCPSM3 assembler compiler are both free and available for download. [95]

The microcontrollers run independently and in parallel to each other within the parallel architecture. The interrupt request (IRQ) based arbiter facilitates efficient communication via the system registers. Each microcontroller only occupies the memory bus during a read or write (R/W) operation. All processing is done in parallel and offline to the memory bus. This significantly relaxes the sequential timing requirements. The designer is free to choose between parallel VHDL code or sequential assembly language which significantly reduces development time. It is often difficult to extend existing VHDL systems. However, this parallel architecture is easily extensible, and the designer can add additional microcontrollers on an ad-hoc basis. This parallel pBlaze architecture is explained in Figure 4.5. The various pBlaze cores and their functions are tabulated in Table 4.1.

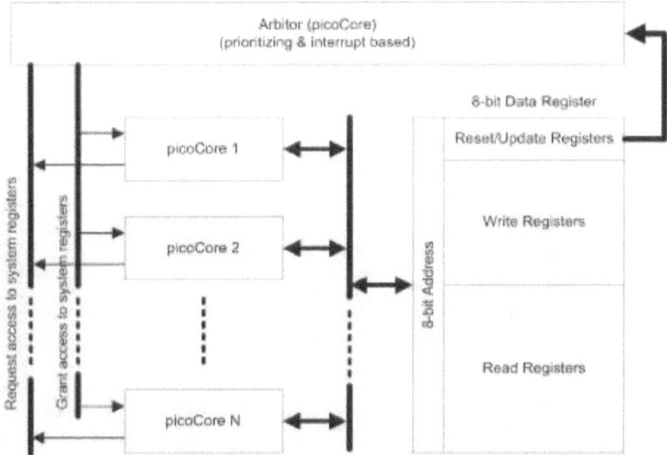

Figure 4.5: PicoBlaze Parallel Architecture. Multiple pBlaze cores run in parallel and communicate via the system registers. The arbiter prioritises and manages bus access. The architecture is entirely IRQ based. Individual cores send IRQs to the arbiter requesting bus access. The arbiter prioritises such requests and grants access by replying with an IRQ to the specific core. The arbiter also notifies the pBlazes of new register data using the same IRQ bus. The upper seven register bytes are special registers used to reset selected modules or to inform the arbiter of new data. The system registers are 8-bits wide. The random access memory (RAM) is split into two identical R/W sections using the most significant bit (MSB) of the address bus. The user can only read from the read register and only write to the write registers. The pBlazes read new data from the write registers. After processing is complete, the data is copied to the read registers. This scheme results in a built-in handshake where the user can verify when and if an operation has completed successfully.

Table 4.1: UCT GPSDO Firmware Blocks

Processor	Address	Function
Arbiter	00 - 07	Manages and prioritises IRQ-based access to the system register bus. The individual pBlaze cores are granted memory access based on interrupts to and from the arbiter.
Chores	various	Responsible for various system chores. Some of these chores include updating the GPSDO status registers, management of the front panel, managing the PSU and system standby.
DAC	0F - 12	Responsible for the PLL frequency steering by interfacing with the TI DAC1220 via an SPI bus. (Section 4.1.3)
AD9512	13 - 18	Responsible for setting up the output clock division and distribution. Interfaces with the Analog Devices AD9512 ia an SPI bus. (Section 4.1.4)
TDC	19 - 2F	Handles the PLL phase detection by interfacing with the ACAM GP2-TDC via an SPI bus. (Section 4.1.2)
RTC	30 - 36	A GPS-synchronous RTC. This module also sets and arms the Epoch Pulse. (Section 4.2.4)
Loop Filter Module	37 - 75	This module includes the sawtooth correction, outlier removal (unimplemented), moving averager, PLL filter and the anti-windup limiter. This module is primarily implemented in VHDL. However, a pBlaze handles the memory management and the sequential control of the VHDL code. (Section 4.2.3).
Phase-Locked Detector	76 - 80	Future use. Not yet implemented.
GPS Rx	81 - AE	Reads and interprets GPS serial strings. (Section 4.1.1)
GPS Tx	81 - AE	Writes serial strings to the GPS. (Section 4.1.1)
Serial Comms	n/a	Manages serial communication with the user. Allows read and write access to the system registers. (Section 4.2.2)

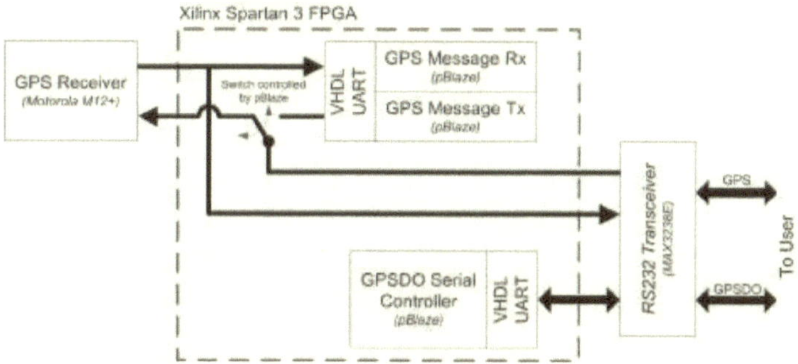

Figure 4.6: UCT GPSDO Serial Communication.

4.2.2 Serial RS-232 Communication

Two two-wire serial interfaces give the user simultaneous and real-time access to both the GPSDO register map and the Motorola M12+ receiver. A single DB9 connector on the GPSDO front panel provides access to both interfaces. See Figure 4.3a. Voltages conform to standard RS-232 specifications. Only a selected few M12+ functions are implemented and available via the GPSDO serial interface. Direct access to the M12+ serial interface gives the user full access to all its functions and ensures compatibility with the existing Motorola WinOncore software[1]. The GPSDO internally communicates with the GPS receiver via two separate pBlazes. Moreover, it intercepts messages from the M12+ before seamlessly passing it on to the user. However, when the GPSDO send messages to the M12+ it momentarily severs user communication on the GPS Tx line. See Figure 4.6.

A dedicated pBlaze handles the GPSDO user communication. Moreover, a serial command string gives the user R/W access to the system registers. Two tilde characters at the start and a line feed and carriage return at the end frame this byte string. A checksum byte ensures data integrity. This string framing and checksum byte is identical to that of the M12+ command string allowing the reuse of VHDL and software code.

---

[1]The Motorola WinOncore provides a GUI with real-time access to the M12+ parameters. It is free and available for download.

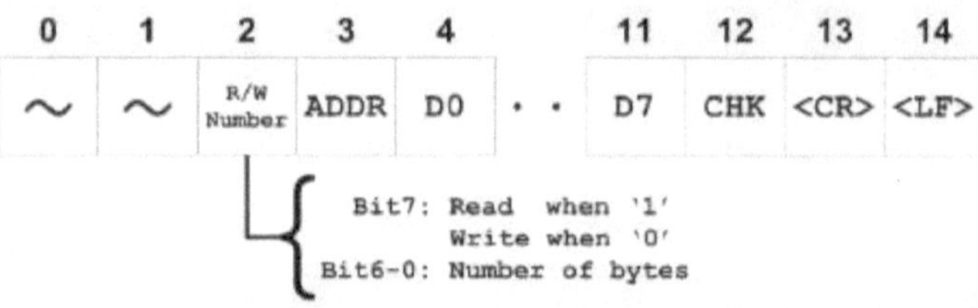

Figure 4.7: UCT GPSDO Serial Command String.

GPSDO Serial Communication Behaviour

The user selects an R/W operation along with the number of bytes to be read or written in byte 2. Byte 3 specifies the register start address. A maximum of 8 data bytes could be read or written at a time. See Figure 4.7.

Each successful message written to the GPSDO is echoed back to the user. Bungled command strings are ignored. No data bytes are included when reading data. The GPSDO will respond with the requested number of bytes starting at the sought address. Data messages are output in the same command string format.

The GPSDO register map contains read-only, write-only, and read-write registers. Writing to a read-only register has no effect. Write-only registers are self-resetting, and it is pointless reading these registers. Data can be written to multiple registers using multiple command strings. However, the GPSDO will only act on register changes once the corresponding update bits located at addresses 03x to 07x are asserted. These update bits make timely and controlled execution possible.

4.2.3 Loop Filter Module

The loop filter module conditions and filters the measured PLL phase error (Section 4.1.2). The output is written to the OCXO-steering DAC (Section 4.1.3). The loop filter module was implemented using VHDL with a pBlaze managing the memory and sequential control. This module includes real-time sawtooth correction, outlier removal (unimplemented), a moving averager, an IIR PLL filter, and an anti-windup limiter. The following section discusses these individual parts as outlined in Figure 4.1.

Sawtooth Correction

The Motorola M12+ GPS receiver outputs an estimate of the sawtooth error expected on the next PPS edge. Refer to Section 2.4.2. Sawtooth correction is performed digitally by subtracting this 8-bit sawtooth error from the measured 16-bit PLL phase error.

Averager

The averager has two modes. It can perform a moving average across an arbitrary window size, or it can calculate the average of N samples and output a value every N seconds dividing both the PLL sampling rate and DAC voltage update rate by a factor of N. This reduced update rate may be advantageous in some situations. Hardin and Yankowski [64] successfully implemented an LSF-based pre-filter in an Rb-based GPSDO. However, the practical PLL time constants for the quartz-based UCT GPSDOs were found to be too short for simple averaging to be of advantage. For this reason, pre-averaging was not used often in the UCT GPSDOs. An LSF-based pre-filter may be more successful but is left for future work.

Outlier Removal

To date, outlier removal remains unimplemented in the VHDL firmware loop filter module. However, a software algorithm was tested for future implementation in VHDL using the provisioned system registers (Section 4.3.1).

2nd-Order IIR-based PLL

A critically damped 2nd-Order PLL is first designed in the s-plane using Egan's method [78]. Then, a stable digital approximation is found by transforming the filter from the s-plane to the z-plane using the bilinear transform [96].

The filter of the form,

$$H[z] = \frac{b_0 + b_1 x^{-1} + b_2 x^{-2}}{1 + a_1 z^{-1} + a_2 z^{-2}} \tag{4.1}$$

is implemented in VHDL using the Direct Form I structure, described by:

$$y[n] = b_0 x[n] + b_1 x[n-1] + b_2 x[n-2] - a_1 x[n-1] - a_2 x[n-2] \tag{4.2}$$

The filter has a bit width of 32, with the input, filter coefficients, and output, each a fixed-point decimal number, $[a.b]$, where $a + b = 32$. The fixed-point arithmetic proved to be challenging. The filter input values and coefficients may vary over a broad range requiring the decimal at different positions. This fixed-point decimal must be handled with care to avoid rounding errors. The Parallel Form structure is less sensitive to coefficient quantisation and would likely have been a better choice [97].

It proved cumbersome to implement and test new and intelligent filters directly in VHDL. Therefore, a software PLL filter module was written (Section 4.3.1). This software filter runs directly on the GPSDO hardware via the serial port bypassing the internal PLL filter. Now, one can rapidly test new and complicated filters before attempting the tedious VHDL coding.

Anti-Windup Limiter

The anti-windup limiter is a simple hard limiter preventing the PLL filter from winding up further than the maximum or minimum possible DAC value. The 32-bit loop filter module is paired to the 20-bit OCXO-steering DAC. Thus, if the filter output is allowed to 'wind up' past 20-bits, it requires an excessive amount of time to 'wind down' again to perform feedback corrections. The anti-windup limiter prevents this. [78]

4.2.4   Epoch Pulse Module

The EP module takes a novel approach to the problem of network-wide time synchronisation. Little is published on GPSDO time synchronisation in the open literature and techniques are often proprietary. Refer to Section 2.5. The EP works much like an alarm clock and can produce a low-jitter GPS synchronised rising edge at an arbitrary user-selected future time. This EP is phase synchronous to the STALO with picosecond level jitter.

A GPSDO steers the STALO's phase towards the mean of the sawtooth corrected high-jitter GPS PPS reference. Thus, the STALO's phase is the best estimate of true UTC at any given moment. The EP module selects the STALO zero-crossing which best coincides with the top of the UTC second.

The EP is implemented in HDL. Refer to the HDL circuitry and associated timing diagram in Figure 4.8. The 10 MHz OCXO output is divided and delayed to allow the high jitter GPS PPS to be unambiguously substituted with a low jitter OCXO rising edge. The real-time clock (RTC) reads the GPS time and increments on each PPS. The RTC asserts `Alarm` one second before reaching the user-programmed alarm time. On every PPS, a state machine, running in parallel, counts the rising edges of the 64 MHz system clock and asserts the `AlarmEn` signal at t_1, approximately 250 ns before it expects the next PPS, exactly one second later. Now, `AlarmEnable` and `Alarm` are AND together producing `Enable`. Thus, `Enable` is asserted 250 ns before the PPS is expected at the pre-programmed alarm time. In the timing diagram, t_3 is the nominal PPS rising edge. However, due to the high jitter, the PPS may arrive anywhere between t_2 and t_4. This peak-to-peak jitter was measured at roughly 45 ns. The 10 MHz OCXO is digitally divided by 10 down to 1 MHz. The goal here is to increase the period, to 1 ms in this case, to make it significantly larger than the PPS jitter. Note that the GPSDO phase-lock's this 1 MHz signal to the PPS reference. Furthermore, the GPSDO PLL offsets the 1 MHz by 65 ns such that it reliably lags the PPS. Now, the first 1 MHz rising edge after the PPS rising edge is a reliable estimate of the true, but 65 ns delayed, GPS time at any given moment. With `Enable` asserted at t_1, PPS will turn high at an uncertain time between t_2 and t_4. The 65 ns PLL offset ensures that the 1 MHz rising edge always arrives after t_4 at t_5. When this happens, the `StartEpoch` rising edge is produced. The `StartEpoch` is then fed back to the RTC, clearing `Alarm`, and therefore, also clearing `StartEpoch` at t_6.

A small HDL tweak generates a low jitter GPS time synchronous PPS pulse train if required. In Figure 4.9 the 10 MHz OCXO phases and EP pulse trains of two GPSDOs were compared. As is expected, the EP error closely tracks that of the 10 MHz OCXO time error. There is a 3.5 ns offset due to part-to-part skew within the EP output buffers. However, careful design can reduce this error and calibration can null it. The 10 MHz time error was measured using the UCT DMTD with femtosecond resolution and peak-to-peak jitter of about 11 ps. The EP error, on the other hand, was measured using an HP53132A with 0.5 ns resolution and peak-to-peak jitter of about 4 ns. (Refer to Appendix C for the HP53132A and UCT DMTD calibration data.) Hence, the EP appears to have much higher jitter. However, one can reasonably expect the EP to have a jitter equal to that of the digital output buffer used to generate it.

In summary, the EP module solves network-wide time synchronisation and relies on two things. Firstly, the PPS signal must be phase-locked to a divided version of the OCXO output. This divided OCXO derivative must have a period significantly larger than the PPS jitter. Secondly, the GPSDO PLL must be offset, by an amount significantly greater than the PPS jitter. This delay ensures that the divided OCXO derivative always lags the PPS. The `StartEpoch` edge will always lag the absolute GPS time by this fixed offset. However, this offset cancels when the relative time error between two or more GPSDOs is measured.

4.3   Graphical User Interface

The GUI enables the real-time monitoring and control of up to three UCT GPSDOs via their respective RS-232 serial ports. Diagnostic data such as the PLL phase errors and DAC steering voltages can also be plot and recorded in real-time. See Figure 4.10 for a screenshot of the GUI and a detailed explanation. The rest of this section discusses the smart PLL filter module comprising of outlier removal, phase-locked detection, and a quick-locking adaptive PLL.

4.3.1   Smart PLL Filter Module

The smart PLL filter module serves as a development environment where sophisticated PLL filters can be rapidly implemented in PC software for verification directly on the GPSDO hardware.

A complex algorithm can first be verified in the high-level software environment before attempting the more tedious and difficult to debug HDL code. This reduces the development time significantly. The smart filter module was defined within a single and separate object class. Thus, in-depth knowledge of the GPSDO GUI is not required to develop and test filter algorithms.

The smart filter bypasses the firmware PLL filter by reading the PLL phase error directly and then writing the filter output directly to the frequency control DAC. At present, this module includes outlier removal, phase-locked detection, and a quick locking adaptive PLL. The quick-locking PLL improves the lock-in time by up to a factor of four when compared to conventional PLLs. See Figure 4.11 for a diagram and a more detailed explanation of the smart PLL filter module.

Future work may include an LSF pre-filter and changing to a Kalman-based control system. Further, the GPS data and measured phase errors of up to three GPSDOs are simultaneously available making it possible to implement common-view

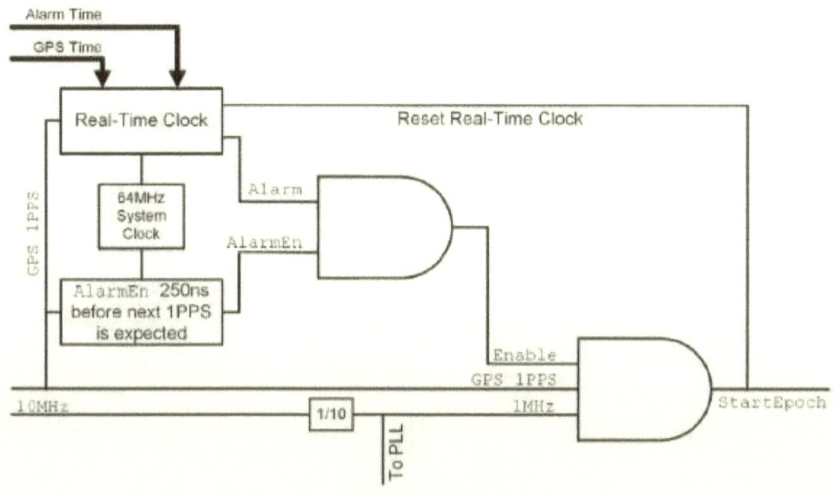

(a) HDL Circuitry

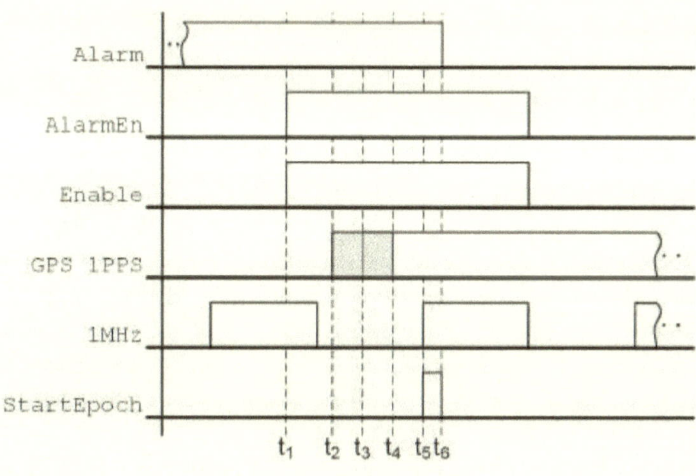

(b) Timing Diagram

Figure 4.8: Epoch Pulse Module

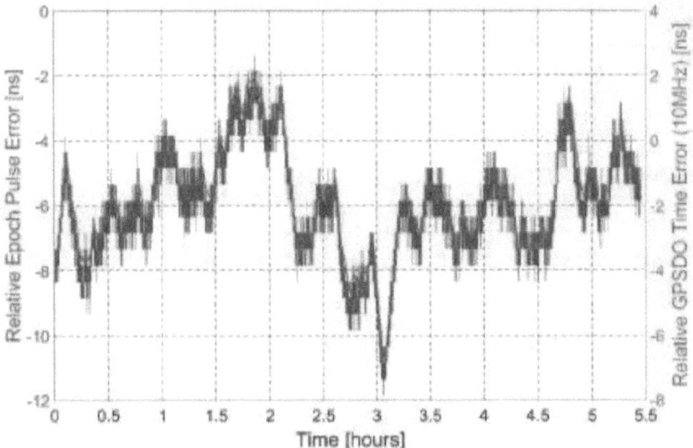

Figure 4.9: Epoch Pulse Performance. The time errors of the 10 MHz OCXO and EP outputs of two GPSDOs are compared. The EP error tracks the 10 MHz OCXO time error closely. The 3.5 ns offset is due to part-to-part skew within the output buffers. However, careful design can reduce this error and calibration can null it.

GPS time transfer.

4.3.2   Outlier Removal

Outlier removal protects against unexpected glitches and is essential in any GPSDO. This interquartile range (IQR) based outlier removal algorithm is simplistic but has reasonable performance. It is quite effective at removing single large outliers but struggles to remove successive outliers that are closely spaced. The IQR is considered a robust statistical measure of centre and spread that is relatively unaffected by the presence of outliers [98]. The mean and standard deviation, on the other hand, are very sensitive to the presence of outliers [98]. Also, the IQR is computationally less intensive than many other robust statistical measures such as the median absolute deviation (MAD).[1]

This algorithm continuously sorts the sawtooth corrected GPSDO phase error values within a moving window in ascending order. Then, the indexes of the 25th, Q1, and the 75th, Q3, percentiles are found. Now, IQR $= Q3 - Q1$. The upper

---

[1]Special thanks to Ulrich Bangert, a 'time-nut', for introducing the author to robust statistics.

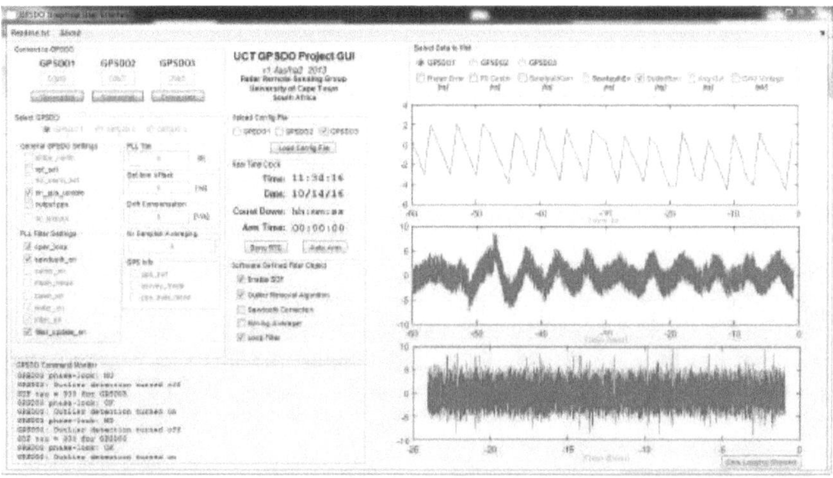

Figure 4.10: UCT GPSDO Graphical User Interface. All three of the UCT GPSDOs can be simultaneously connected to the GUI via the `Connect to GPSDO` panel. The `Select GPSDO` panel contains various checkboxes, each representing a bit setting in the GPSDO registers. For example, the GPSDO could be turned open-loop by checking the `open_loop` box. The `GPSDO Command Monitor` window displays and logs to file all the GPSDO setting changes during a session. The time-stamped log file could later be used to interpret recorded phase error data. The `Upload Config File` facility allows the user to upload text files containing custom commands to one or multiple the GPSDOs. This is a powerful tool which allows the user to R/W non-standard commands to/from any number of the GPSDO registers. The `Real-Time Clock` panel displays the current GPS time and permits the user to arm the EP to an arbitrary future time. The `Auto Arm` button automatically arms the EP five seconds in the future. The `Software Defined Filter Object` represents the smart PLL filter module described in Section 4.3.1. Various GPSDO parameters are plotted in real-time within the `Select Data to Plot` panel. It contains three axis which respectively represents a minute, an hour and a day worth of data. Data logging could be stopped and started arbitrarily.

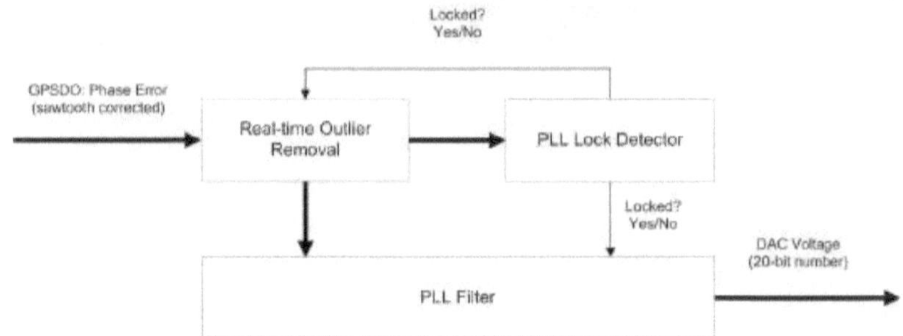

Figure 4.11: Diagram of the Smart PLL Filter Module. The sawtooth corrected PLL phase error is input to the outlier detection module. This module, only active while the PLL is locked, discards outliers and passes non-outliers to the PLL filter module. The PLL filter first locks the PLL using a short time constant of $\tau_{PLL} = 75\,\mathrm{s}$. Then, when PLL lock is detected, the PLL filter time constant, τ_{PLL}, is ramped until it reaches the final τ_{PLL} value. This method of ramping the filter time constant while keeping constant the damping of the feedback loop improved the PLL lock time by a factor of four over a conventional fixed τ_{PLL} PLL filter.

and lower bounds are defined as, $O_{lower} = Q1 - 1.5\mathrm{IQR}$, and $O_{upper} = Q3 + 1.5\mathrm{IQR}$, respectively. Input values that are within these limits pass through to the output directly. Out of bound inputs are discarded with the output held unaltered. See Figure 4.12 for an example of where the outlier removal is applied to real data.

Ideally, the outlier algorithm should be tailored to the PPS error characteristics, but this is left to future work.

4.3.3 Phase-Locked Detection

Phase-locked indication is crucial for GPSDO monitoring. However, the literature is particularly sparse on this topic. A GPSDO can be considered phase-locked when its oscillator's phase closely follows that of the GPS PPS reference. Since frequency is the derivate of phase, the GPSDO's output frequency will then also be a closely matched multiple of the GPS PPS reference. Thus, when phase-locked, both the instantaneous PLL phase error, as well as its slope, will be within certain bounds and centred around zero.

This phase-locked detection algorithm relies on two criteria. Instantaneous absolute PLL phase error, and the average PLL phase error over a fixed moving window.

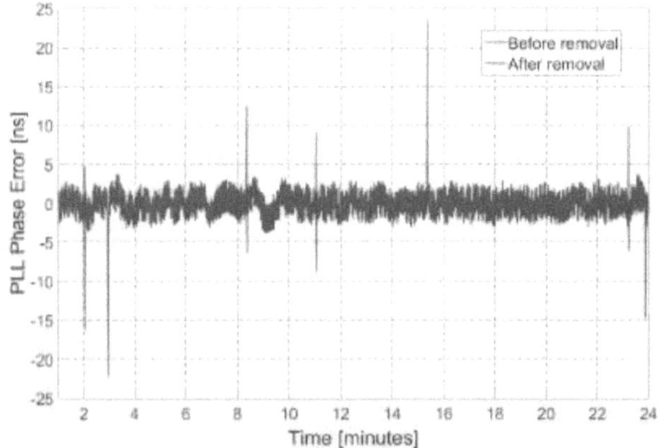

Figure 4.12: Real-Time Outlier Removal. This simple algorithm is quite effective at removing single large outliers. However, it struggles to remove successive outliers that are closely spaced.

Thresholds with hysteresis are assigned to both these measures. These thresholds are dependent on the characteristics of the GPS PPS reference and were determined empirically for the sawtooth corrected PLL errors produced by the UCT GPSDOs.

Other measures, such as phase slope or ADEV, across a moving window, were also considered. However, both of these criteria are computationally more intensive in comparison to the average slope. Also, the ADEV proved inconvenient since it is highly dependent on the window size. Conversely, the average phase error is largely unaffected by the window size allowing its threshold and the window size to be adjusted independently.

This algorithm assumes that the PLL is of the second-order and critically damped. It follows that all locking transients are of similar shape but scaled by the filter time constant. The algorithm is required to produce comparable results independent of the filter time constant. Hence, the window size is set equal to the PLL filter time constant in seconds. This window size adjustment ensures reliable performance independent of the filter time constant. See Figure 4.13 for an example of the algorithm output when applied to real sawtooth corrected PLL phase error data.

Through trial and error, the above algorithm was empirically matched to the

PPS error characteristic. Ideally, the phase-locked detection algorithm should auto adjust to the reference input, but this is left to future work.

4.3.4 Quick-locking Adaptive PLL

Excessive lock-in times caused by long PLL time constants proved impractical during the NetRAD trials documented in Chapter 6. Subsequently, the UCT GPSDOs were tuned to a suboptimal $\tau_{PLL} = 150\,$s for the duration of the trials. Moreover, the critically damped response was found to take roughly 3τ to achieve reasonable lock when assuming no cycle slipping. Hence, the UCT GPSDO requires at best 50 minutes to lock when adjusted to its optimal $\tau_{PLL} = 1000\,$s. However, the short $\pm 500\,$ns phase detector range makes the UCT GPSDO prone to cycle slipping at long time constants and PLL lock-in could take significantly longer. To solve this problem, a novel quick-locking 2nd-order PLL was designed. This adaptive filter reduces the lock-in time by a factor of four. It relies on the ability to both detect phase-lock (see Section 4.3.3) and change the bandwidth in real-time without altering the PLL damping.

One may start by determining the five coefficients of the PLL filter for the critically damped case using Egan's method [78]. Coefficients are calculated for various τ_{PLL} between 75 seconds and 2000 seconds.[1] Then, a curve of the form $y(x) = ax^b + c$ is fitted to each set of coefficients. The resulting equations describe the five filter coefficients as independent functions of τ_{PLL} for the critically damped case. It is now possible to arbitrarily adjust τ_{PLL} without altering the damping of the feedback loop.

To achieve rapid phase-lock, the UCT GPSDO is tuned to $\tau_{PLL} = 75\,$s which is the shortest practical time constant. Then, when phase-lock is detected, τ_{PLL} is gradually ramped by changing the filter coefficients in real-time. Transient effects are avoided by changing the coefficients gradually enough and only when the average PLL phase error has a slope of zero. The time constant, τ_{PLL}, can be ramped linearly but it was found that it could be ramped quicker for lower τ_{PLL} values without inducing any significant transients. Thus, higher GPSDO stability is reached quicker by ramping τ_{PLL} according to a cubic polynomial. The time constant, τ_{PLL}, is ramped to reach the final value in $0.6\tau_{PLL}$. The UCT GPSDO can now reliably

---

[1]An underdamped response can be achieved similarly. However, this text is limited to the critically damped case.

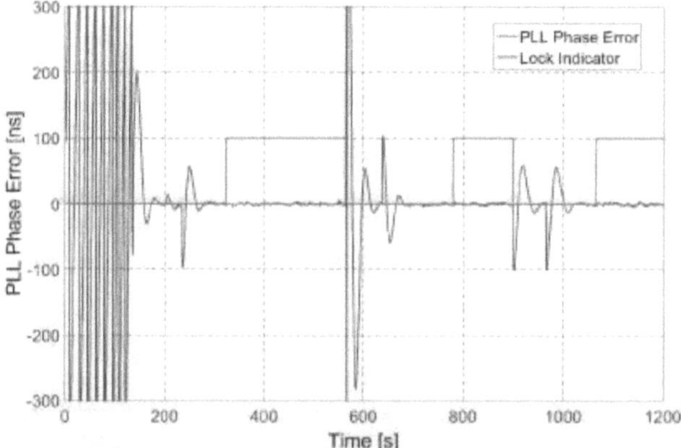

Figure 4.13: Real-Time Phase-Locked Detection. This plot shows the output of the phase-locked detection algorithm in response to real PLL phase error data. The dataset purposely contains numerous locking transients. Note how the algorithm waits for locking transient to decay before it classifies the PLL as locked.

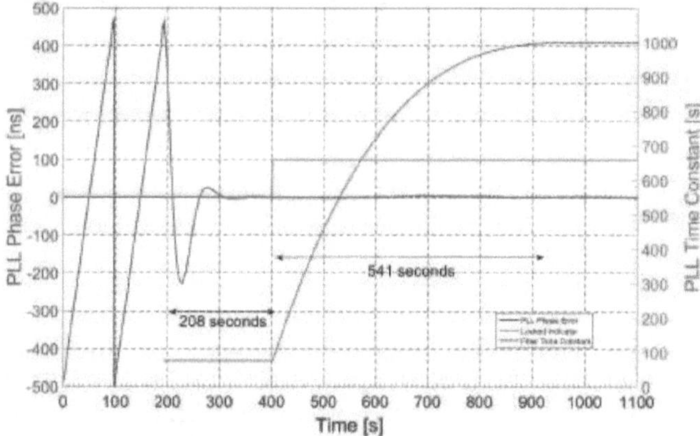

Figure 4.14: Quick-Lock Adaptive PLL. A real-world demonstration of the quick locking PLL. The PLL, with $\tau_{PLL} = 75\,$s, is turned on at around $200\,$s. Then, $208\,$s later, phase-lock is detected. Now, τ_{PLL} is ramped while keeping the damping constant, following the green curve, until the target $\tau_{PLL} = 1000\,$s is reached $541\,$s later. A $\tau_{PLL} = 1000\,$s is reached in $749\,$s ($0.75\tau_{PLL}$). Thus, it improves the lock-in time by a factor of four over the $3\tau_{PLL}$ lock-in time of conventional PLLs.

and repeatedly phase-lock to $\tau_{PLL} = 1000\,\text{s}$ in 12.5 minutes. See Figure 4.14 for a demonstration of the quick-locking PLL filter.

Today, this algorithm is successfully implemented in the newer NeXtRAD system [33]. It is particularly convenient for NeXtRAD measurements because the optimal synchronisation performance is often not required for initial test transmissions. Test transmissions can start in under 3.5 minutes as soon as the PLL locks to $\tau_{PLL} = 75\,\text{s}$. From here on, the synchronisation performance gradually improves as τ_{PLL} is ramped. After 12.5 minutes, the PLL is locked to $\tau_{PLL} = 1000\,\text{s}$, and full performance radar recordings can commence immediately after completing the test transmissions.

Other PLL architectures exist such as frequency assisted phase-locked loops (F-PLLs) where an FLL assists with phase-locking continuously [80]. Baojian, Dehai, and Dazhi [30] took a novel two-step approach to phase-locking. First, they implemented an FLL to frequency-lock the GPSDO. Then, they used a phase shifter to adjust independently the GPSDO phase to achieve phase-lock. It is left as future work to compare the performance of these designs to the quick-locking PLL described above. It would also be desirable to real-time adjust the PLL damping factor, but this is also left to future work.

4.4 Conclusion

This chapter described the hardware, firmware, and GUI design of the low-cost quartz UCT GPSDO. The goal of designing an open, versatile, and accessible GPSDO to investigate network radar synchronisation was achieved. Moreover, three identical UCT GPSDOs were built, calibrated in Chapter 5, and later used to synchronise NetRAD in Chapter 6. They solved three practical difficulties. Firstly, they are capable of network-wide time synchronisation that is up to an order of magnitude more accurate than what is possible with the raw GPS PPS signals. (See Section 5.2) Secondly, a quick-locking adaptive PLL was implemented that can phase-lock up to four times quicker than a conventional PLL. Thirdly, the GUI includes a development environment where sophisticated and intelligent control loops can be rapidly developed on a PC for verification directly on the GPSDO hardware.

The Motorola M12+ GPS timing receiver paired with the Oscilloquartz 8788 OCXO as STALO proved to be a good choice. Both devices offer exceptional performance in the lower price range. The TDC-GP2 with its 65 ps time resolution performed well and as expected. It has the advantage of measuring time directly and does not require excessive digital division ratios. However, the TDC-GP2 limits the measurement range to ±500 ns and the HDL code controlling and gating it is somewhat complicated. The 20-bit SPI controlled DAC1220 performed well steering the OCXO. However, due to project time constraints, the supporting circuitry never made it past the prototyping stage. Hence, it is overcomplicated, and leaves some room for improvement. Ideally, the circuitry should limit noise and frequency drift to within the 20-bit LSB resolution. The current circuitry has likely only achieved stability to within the TDC LSB resolution. However, there was never an opportunity to calibrate the steering circuitry properly. Nevertheless, it proved adequate for the radar measurements in Chapter 6. Refer to Section 4.1.3 for the recommended steering-circuitry improvements. The clock distribution is versatile with various outputs available in various digital formats. However, due to time constraints and imminent NetRAD trials, provisioned common-base BJT amplifiers were never populated. Instead, the noisy ERA-6SM+ buffers, meant for testing only, were used. These ERA-6SM+ buffers increased the output phase noise by 15 dBc/Hz to 20 dBc/Hz. Hence, it is highly recommended to populate the common-base buffers for future experiments.

The parallel multi-pBlaze architecture is versatile and easily extensible where new modules are added in an ad-hoc fashion. The entire GPSDO firmware is implemented using this architecture. The firmware is fully autonomous and does not require intervention from external software. The serial communication is robust and has a near identical protocol to that of the M12+. It allows the user real-time access to the low-level GPSDO parameters. The PLL filter includes real-time sawtooth correction, an optional prefilter, outlier removal (not fully implemented), and an IIR filter. The IIR filter coefficients can be adjusted arbitrarily and in real time. The Epoch Module solves the problem of network-wide time synchronisation. This module produces an accurate GPS synchronised pulse at an arbitrary user-selected future time. It offers the best estimate of true UTC at any given moment, and it is up to an order of magnitude more accurate than the raw GPS PPS output. (See Section 5.2)

The GPSDO GUI enables real-time monitoring and control of up to three GPSDOs via their respective RS-232. Diagnostic data such as PLL phase errors and DAC steering voltages can be plot and recorded in real-time. This GUI includes a smart

PLL filter module. This module serves as a development environment where sophisticated PLL filters can be rapidly developed on a PC for verification directly on the GPSDO hardware. This environment was used to implement outlier removal, phase locked detection, and ultimately a quick-locking adaptive PLL. The phase-locked detection functions reliably and is insensitive to the PLL time constant. The quick-locking adaptive PLL reaches phase lock up to four times faster than a conventional PLL. This PLL relies on the ability to both detect phase-lock and change its bandwidth in real-time without altering the PLL damping. Today, this algorithm is implemented successfully in the newer NeXtRAD system [33]. It is also possible to implement Kalman and common-view based but this is left to future work.

Chapter 5

Calibration of the UCT GPSDOs

This chapter presents the synchronisation performance of the UCT GPSDO. The relative time (phase), frequency, frequency stability and phase noise performance is measured at a zero baseline. The purpose is to calibrate the baseline performance and to verify the correct operation of the UCT GPSDOs. It is shown simultaneously how time-domain clock characterisation is used to diagnose and verify GPSDO performance. Both the hold-over and phase-locked performance are calibrated for the case of multi-channel one-way GPS time transfer. These zero-baseline measurements represent the best possible performance under comparatively ideal circumstances. The results are used to predict bistatic radar performance which is later compared to the real bistatic radar performance recorded in Chapter 6. It is expected that the zero-baseline results will deteriorate somewhat with increasing baselines [3].

The testbench was set up carefully to minimise time biases and to ensure good GPSDO and antenna temperature tracking. At the time, the GPSDO were well-aged for more than a month. The GPS PPS performance is measured using two dual-channel HP 53131A TICs, whereas the sinusoidal 10 MHz GPSDO time (phase) is measured using the UCT DMTD system. This testbench is calibrated in Section C.2.

The GPS PPS stability binds the long-term GPSDO stability. Therefore, the relative timing performance of the built-in Motorola M12+ receivers is verified. The inherent combined antenna-receiver time biases were calibrated to be less than 4 ns. Published data often report on the absolute, and or time-averaged performance versus UTC, and less frequently on the GPS-to-GPS performance across baselines of few kilometres. Refer to Section 2.4.4. The sawtooth corrected PPS signal was measured to have a time accuracy of about 28 ns translating to a bistatic baseline error of 8.3 m.

The GPSDO hold-over performance was measured for a few reasons. First, it is demonstrated that a GPSDO can self-calibrate [68] the optimal PLL time constant by making a free-running GPSDO-UTC comparison using its built-in phase detection circuitry. The optimal time constant for the quartz-based UCT GPSDO was estimated at 1000 s. Secondly, a free-running frequency stability plot is a convenient way to verify the proper function of the GPSDO steering circuitry and clock distribution. Lastly, there may be a loss of GPS coverage due to physical obstruction, equipment failure, loss of power, or deliberate or accidental signal jamming. GPSDO hold-over performance is vital in such GPS 'denied' environments. As expected, the quartz-based GPSDOs are not suitable for bistatic time synchronisation during extended hold-over periods. Moreover, the UCT GPSDOs drift by as much as a few microseconds per day. Nonetheless, it can still provide a bistatic Doppler accuracy of a few Hertz at radar frequencies of a few gigahertz during a day of hold-over.

The closed-loop GPSDO performance is measured at various PLL filter time constants. It is shown that the time and frequency performance improves for decreasing PLL bandwidths and that optimal performance is achieved at a PLL time constant of a 1000 s. Moreover, at this optimal time adjustment, it achieves a time accuracy of ± 3 ns which equates to a bistatic baseline error of ± 0.9 m or a carrier phase variation of about $\pm 2600°$ at 2.4 GHz. As is expected from a properly functioning GPSDOs, the phase-locked UCT GPSDOs achieved atomic level long-term frequency accuracy (10^{-13}), and frequency drifts rates (10^{-14}). However, the performance is limited by the short-term frequency accuracy of a few parts in 10^{-10}.

Finally, the GPSDO phase noise is measured. The phase noise level of the 100 MHz LVPECL GPSDO clock was measured to be -70 dBc/Hz at 1 Hz where the Flicker FM noise drops off to -115 dBc/Hz at 100 Hz. This performance is sufficient for the bistatic NetRAD measurements in Chapter 6 where the data will be used to assess the bistatic phase noise response. However, due to a poorly chosen clock buffer, the phase noise level is between 15 dBc/Hz to 20 dBc/Hz higher than expected. Therefore, it is strongly recommended that the discrete common-base BJT buffer amplifier is populated for future measurements.

5.1   Measurement Setup

In this chapter, the zero-baseline relative time differences between the M12+ GPS PPS edges, the GPSDOs during hold-over, and the phase-locked GPSDOs are measured at one-second intervals. From these time errors, the fractional frequencies, and frequency stability is computed. This section describes the measurements setup.

The bistatic measurements in Chapter 6 used GPSDO3 at the transmitting node and GPSDO1 and GPSDO2 for the two passive bistatic receiver nodes. Thus, to assist in comparison, this chapter mainly presents the relative results for GPSDO31. The performance of GPSDO31 was found to be nearly identical to GPSDO32.

The GPS receivers were configured for multi-channel one-way time transfer. Moreover, the three Motorola Timing2000 GPS antennas were mounted on the roof with an unobstructed view of the sky. The antennas were co-located to minimise relative positional errors and to ensure good temperature tracking. The propagation delay of the respective antenna cables was measured and programmed into each M12+ receiver. The auto-survey function of one of the M12+ receivers was used to survey the antenna coordinates. Then, all three receivers were programmed to the same position and put into 'position hold' mode. Hence, the time bias caused by the positional inaccuracy is common to all. The M12+ receivers track up to twelve satellites simultaneously.

The working student laboratory is non-ideal and poorly airconditioned with a daily ambient temperature variation of up to 20°C. Therefore, the GPSDOs were deliberately co-located to improve temperature tracking. The enclosure temperatures were confirmed to track to within 1°C, albeit having constant offsets of a few degrees Celsius.[1] All measurement cables were kept short and the cable lengths matched to within 1 cm keeping the approximate cable time biases below 50 ps.

The relative PPS timing differences between the Motorola M12+ receivers were measured simultaneously using the two dual-channel HP 53131A TICs. Moreover, the one M12+ receiver served as the reference to which the other two M12+ receivers were compared. The sawtooth data for each receiver were recorded simultaneously, and sawtooth correction was applied during post-processing. Refer to Section C.2.1 for the time interval counter (TIC) calibration data.

The relative GPSDO time (phase) error was measured simultaneously using the four channels of the UCT DMTD system. Moreover, the zero-crossing time differ-

---

[1]The M12+ receiver's onboard oscillator was used to log the enclosure temperature.

ence between the 10 MHz sinusoidal OCXO outputs are compared at a DMTD beat frequency of 1 Hz. The one GPSDO was used as the reference to which the other two GPSDOs were compared. During the hold-over measurements, the GPS-OCXO time difference was recorded simultaneously using the built-in GPSDO TDC. Thus, the relative phase and UTC time differences were recorded concurrently. At the time of the measurements, the GPSDOs were well-aged for more than a month. The GPSDO sawtooth correction was active during all measurements. Refer to Section C.2.2 for the UCT DMTD calibration data.

Note that the GPSDO timekeeping capability was measured by comparing the zero-crossing time differences of the 10 MHz clock outputs. The UCT GPSDO's EP function performs real-time network-wide time synchronisation to the same accuracy but with a GPSDO-to-GPSDO skew of about 4 ns. See Section 4.2.4.

The frequency stability plots (ADEV, MDEV and TDEV) were computed using the Stable32[1] software suit [99] and plotted using Matlab. Throughout this chapter, the convention is followed where GPS32 means the relative error between GPS3 and GPS2 such that GPS32 = GPS3 − GPS2.

5.2 Relative GPS PPS Performance

In this section, the PPS timing performance of the built-in Motorola M12+ GPS receivers is calibrated and verified.

A 48-hour plot of the relative M12+ PPS time error is given in Figure 5.1. The sawtooth corrected data are displayed in black. Note that the plots for GPS21 and GPS31 were deliberately offset for clarity. The PPS performance is nearly identical in all three cases.

The mean receiver time offsets range between -2.7 ns and 3.1 ns. Refer to Table 5.1. These mean offsets stayed unchanged to within 1 ns through receiver power cycles. Also, the mean error is unaffected by sawtooth correction. This suggests that the relative zero-range time bias could be nulled to within about 1 ns contributing about 0.3 m to the bistatic range error. However, a possible firmware bug was once detected where the mean offsets kept changing by up to 6 ns before and after power cycles. However, this was resolved by resetting the receivers to their default settings.

---

[1]The Stable32 software suite was donated to the IEEE UFFC in 2018. The software is now available as a free download.

The approximate peak-to-peak and RMS timing errors of the raw PPS signals are 80 ns and 13 ns, respectively. After sawtooth correction, the peak-to-peak and RMS timing errors improved to roughly 28 ns and 3 ns. A histogram comparison of the raw and sawtooth corrected GPS31 timing errors is displayed in Figure 5.2. Both plots are roughly Gaussian shaped with some visible quantisation effects. The sawtooth removal improves the relative peak-to-peak and the RMS timing errors by a factor of 3 and 5, respectively.

The ODEV for the GPS31 PPS time error is plotted in Figure 5.3. The slopes for both the raw and sawtooth corrected cases are constant at -1. An MDEV analysis confirmed the noise type to be White PM noise. Sawtooth removal improved the ODEV by a factor of five.

The performance of the three M12+ GPS receivers appears to be consistent with what is reported in the literature [60]. See Section 2.4.4. For network-wide time synchronisation, the raw PPS signals will limit the time accuracy to about 80 ns or a bistatic baseline error of roughly 24 m. If real-time sawtooth correction is used, the time accuracy improves to about 28 ns or a bistatic baseline error of roughly 8.4 m. The timing noise is mostly Gaussian which suggests that the time accuracy, and frequency stability, can be improved through GPSDO PLL filtering. The mean time offsets (time precision) appear to be constant and can be nulled through calibration. However, surveying inaccuracies [31] and antenna temperature gradients [61] are expected to introduce a few nanosecond bias under field conditions.

5.3 Relative hold-over Performance

In this section, the hold-over performance of the UCT GPSDO is measured. Thus, the combined hold-over performance of the built-in Oscilloquartz 8788 OCXO and the oscillator steering circuitry is measured. The UCT GPSDOs perform no hold-over compensation. First, the well-aged GPSDOs were kept GPS-locked for a few hours. Then, at the start of the measurement, they were switched to free-running mode. In free-running mode, the OCXO steering circuitry holds the last DAC value. Both the GPSDO-UTC and the GPSDO-to-GPSDO time errors are measured concurrently. Finally, the fractional frequency and frequency stability are computed.

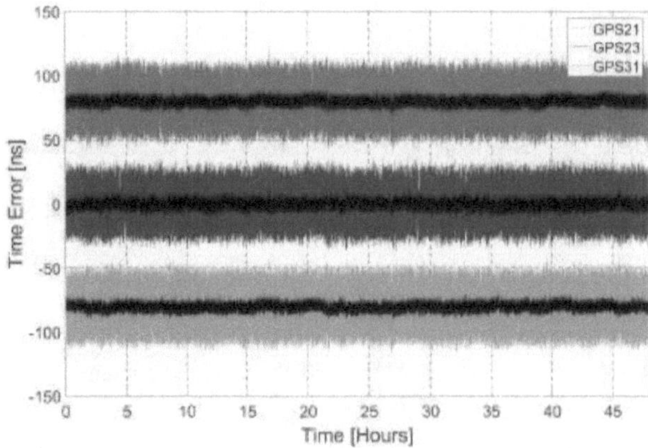

Figure 5.1: Relative M12+ GPS PPS Time Error (48 Hours). The sawtooth corrected data are displayed in black. The mean time offsets were removed before GPS21 and GPS31 were offset for clarity.

Table 5.1: Relative M12+ GPS PPS Time Error Statistics

Description[a]	Mean	Pk-Pkr	RMS
	[ns]	[ns]	[ns]
GPS21	0.379	80	12.8
GPS23	3.113	73.5	12.7
GPS31	-2.724	73.5	12.8
GPS21*	0.379	27.5	2.8
GPS23*	3.113	24.5	2.7
GPS31*	-2.724	24.5	2.8

[a]Sawtooth corrected data are denoted with an asterisk.

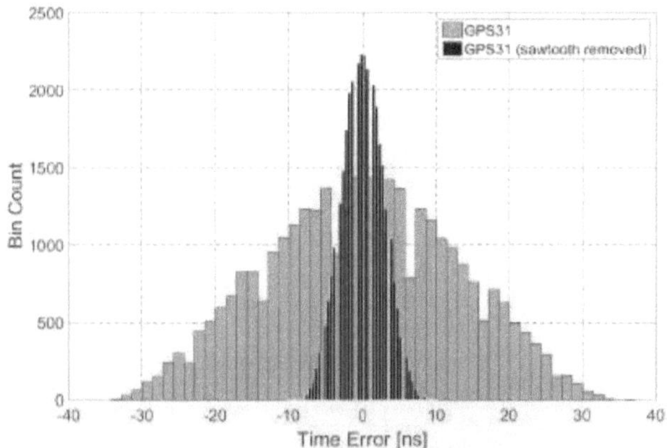

Figure 5.2: Histogram of the Relative M12+ GPS PPS Error. Both errors appear to be roughly Gaussian distributed with some quantisation effects. Sawtooth removal reduces the peak-to-peak and RMS time errors, by a factor 3 and 5, respectively. The mean time offsets were removed.

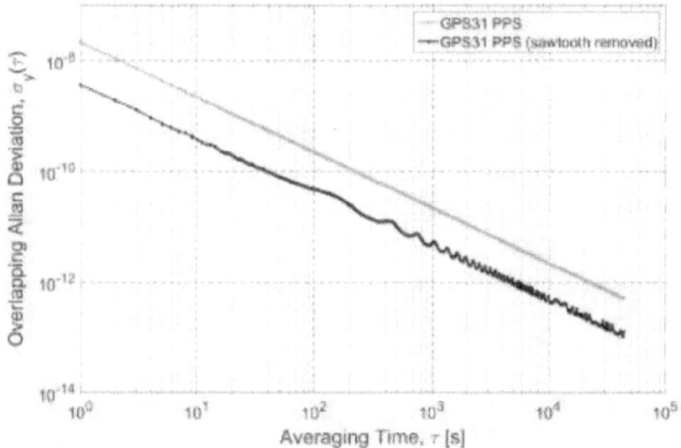

Figure 5.3: Overlapping Allan Deviation of the Relative M12+ GPS PPS Error. The slope of -1 is indicative of White PM noise (confirmed by MDEV). Sawtooth removal improves the ODEV by a factor of 5.

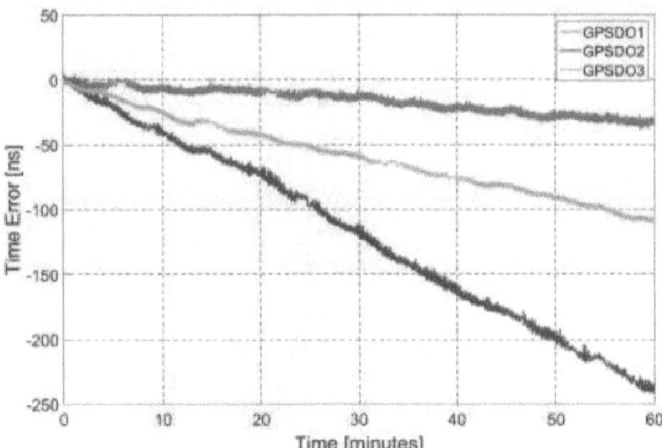

Figure 5.4: Free-Running UCT GPSDOs compared to UTC (1 hour). The hold-over performance versus UTC was recorded using the GPSDO's built-in TDC. The GPSDOs were phase-locked to GPS before turning them open-loop at the start of the recording. Each plot has a constant initial slope indicative of a frequency offset ranging between -6.7×10^{-11} and -9×10^{-12}.

5.3.1 Time (Phase) Error

The free-running UCT GPSDOs are compared to UTC in Figure 5.5. Note the remaining residual high-frequency noise after sawtooth correction. At the start of the recording, all three GPSDOs were phase-locked to GPS. However, immediately after switching to open-loop mode, the GPSDOs started to drift off, albeit along constant slopes. These constant slopes represent initial frequency offsets ranging between -9×10^{-12} and -6.7×10^{-11} with negligible drift within the first hour. Thus, confirming the initial frequency offset as the dominant source of time error during medium-term hold-over. Nonetheless, the bistatic Doppler errors induced by such frequency offsets are negligible at radar frequencies. However, after an hour of hold-over, the GPSDO may drift off by hundreds of nanoseconds, where every hundred nanoseconds relate to a bistatic baseline error of 30 m. As expected, such quartz OCXOs cannot provide sufficient hold-over for more than a few minutes if a bistatic range accuracy of a few metres or less are required.

The relative GPSDO hold-over performance for a period of 11 days is given in Figure 5.5. It is evident that GPSDO3 drifted more than the other two. In Sec-

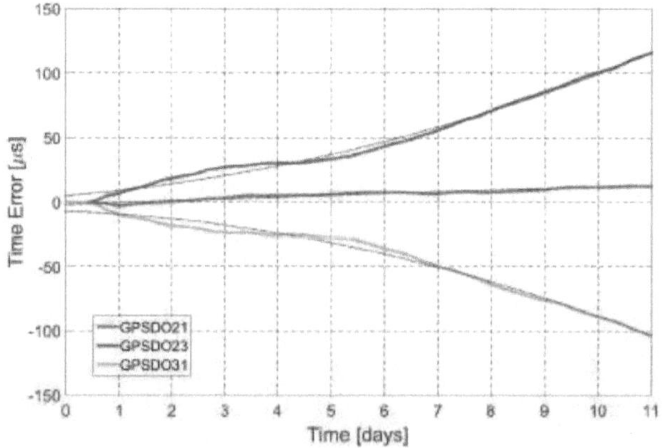

Figure 5.5: Relative UCT GPSDO Hold-Over Performance (11 days). The black dotted lines represent 2nd-order polynomial curve fits. It is not immediately clear why there was an abrupt behavioural change after about half a day.

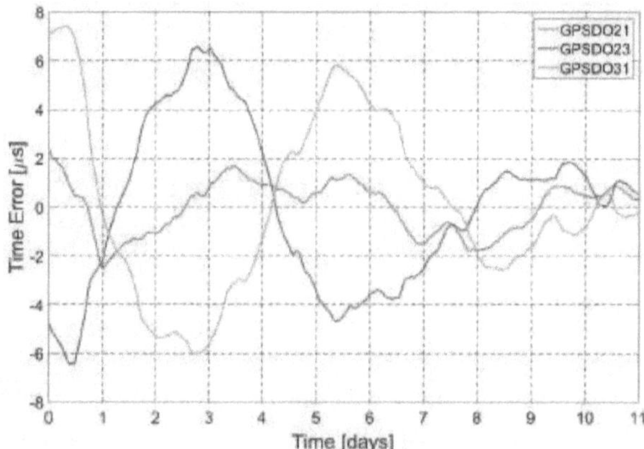

Figure 5.6: Residual Relative UCT GPSDO Hold-Over Performance (11 days). The respective 2nd-order polynomial fits were subtracted from the plots in Figure 5.5 to produce the residual relative time errors.

tion A.1, it was described how a 2nd-order polynomial might model the deterministic oscillator time error. To this end, a 2nd-order polynomial is fit to each plot and indicated by the black dotted lines. However, the slow varying modulation effects cause the polynomial fits to be fairly inaccurate. It is suspected that the modulation may originate from the OCXO steering circuitry rather than the OCXO itself. Nonetheless, Figure 5.6 show the respective residual time differences after subtracting each polynomial fit. It follows that a 2nd-order polynomial based hold-over algorithm may significantly improve the free-running time error. However, even with excellent hold-over compensation, the time error is expected to still be on the order of a few microseconds per day.

5.3.2 Frequency Error

The free-running fractional frequency of GPSDO31 over one day is given in Figure 5.7. The black trace is a one-minute moving average to better show the frequency drift over time. The plot shows a peak-to-peak frequency noise of about 10×10^{-10} which improved to around 2×10^{-10} after the 8th day. It is not clear what caused this improvement. However, the frequency noise is about an order of magnitude higher than what can be expected from this class of GPSDO. The long-term peak-to-peak variation for the one-minute averaged plot is on the order of 3×10^{-10}. Hence, the UCT GPSDOs are expected to maintain an average bistatic Doppler accuracy of a few Hertz at radar frequencies of a few Gigahertz over short baselines over 24 hours. However, in this case, the bistatic hold-over performance may be limited by the high short-term frequency of a few parts in 10^{-10}. NetRAD requires a frequency accuracy of below 2.1×10^{-10}. See Section 3.6.3.

5.3.3 Frequency Stability

This section presents the relative frequency stability of the UCT GPSDOs during hold-over. The purpose is to calibrate the hold-over performance, verify the GPSDO operation, and to determine the optimal GPSDO PLL time constant. The optimal GPSDO PLL time constant is at the cross-over point between the frequency stability of the internal OCXO and UTC.

In Figure 5.8, the free-running GPSDO frequency stability versus UTC is considered. Here, internal GPSDO TDC directly compares the open-loop OCXO stability to UTC. For comparison, the sawtooth corrected GPS PPS plot is shown in

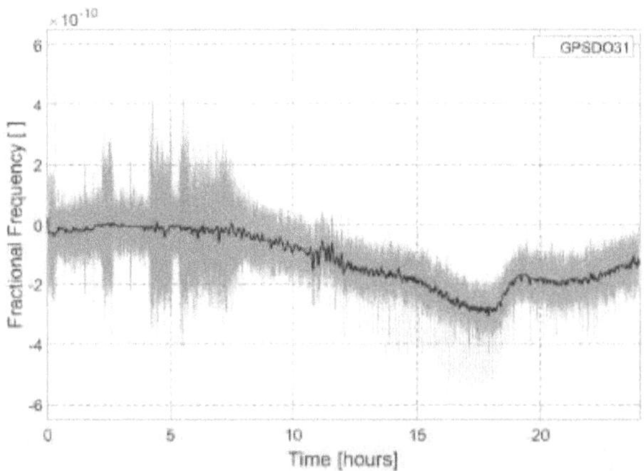

Figure 5.7: Free-Running Fractional Frequency for GPSDO31 (1 day). The green plot show the real-time fractional frequency. The black trace is a one-minute moving average.

black. As expected, the short-term OCXO frequency stability is masked by the high-frequency noise of the GPS PPS signal. The $\sigma_y(\tau = 1)$ discrepancy indicates differing noise levels within the phase detection circuitry of the GPSDOs which are likely due to the prototype manufacturing. At around 1000 seconds, the OCXOs start to drift off while the GPS PPS signal keeps improving for increasing averaging times. Hence, the optimal GPSDO PLL time constant appears to be near 1000 seconds. Judging by the $\sigma_y(\tau)$ plot for the sawtooth corrected PPS reference one may guess that the UCT GPSDO may achieve a closed loop $\sigma_y(\tau = 1000) \approx 5 \times 10^{-12}$.

Plots of the relative ODEV for free-running UCT GPSDOs are given in Figure 5.9. The GPSDO21 performance is as expected, where White PM, Flicker PM and Flicker FM noise are present up until about 10 seconds after which Random Walk FM noise starts to dominate. However, a lower noise OCXO buffer amplifier (see Section 4.1.4 and Section 5.5) is expected to result in a factor of ten improvement of both the $\sigma_y(\tau = 1)$ and short-term frequency accuracy, lowering the levels to a few parts in 10^{-12} and 10^{-11}, respectively. It is not clear why $\sigma_y(\tau = 1)$ of GPSDO31 is 1.5 times higher than the rest. This measurement compared the three GPSDO simultaneously, and instability that is originating from one GPSDO should appear in at least two plots. The plots of GPSDO31 and GPSDO32 show a hump between 4 seconds and 300 seconds caused by excessive Flicker FM[1]. This hump suggests some instability in the GPSDO3 OCXO steering circuity explaining its poorer hold-over.

This section demonstrated that hold-over frequency stability is a powerful diagnostic tool. The short-term stability of the UCT GPSDOs can be improved by populating the low-noise BJT OCXO buffer amplifiers, and GPSDO3 appears to have some steering instability.

5.4 Relative Closed-Loop Performance

In this section, the UCT GPSDO performance is measured at various PLL time constants. The purpose is to verify that the GPSDOs function correctly, to confirm that an appropriate PLL time constant is selected, and to calibrate the GPSDOs for comparison to the bistatic radar measurements in Chapter 6. The relative time, frequency and frequency stability are presented. Note that throughout this text, the UCT GPSDO PLL is tuned critically damped. Refer to Section 4.2.3 and Section 4.3.1 for more on the filter design and Figure 4.14 for a plot of the critically damped locking transient.

5.4.1 Time (Phase) Error

This section presents the phase-locked UCT GPSDO time (phase) error at various time constants. An eight-hour plot of the GPSDO31 time error is given in Figure 5.10. This plot compares the phase-locked timing performance for $\tau_{PLL} = 150\,$s versus $\tau_{PLL} = 1000\,$s. The mean time offsets were removed. The peak-to-peak timing error for $\tau_{PLL} = 150\,$s is roughly $\pm 5.5\,$ns equating to $\pm 20°$ at the $10\,$MHz OCXO frequency. The reduced loop bandwidth for $\tau_{PLL} = 1000\,$s improves the peak-to-peak timing error to approximately $\pm 3\,$ns which equates to $\pm 10.8°$ at $10\,$MHz. The histograms of the two plots are given in Figure 5.11. It is anticipated that the noise will become more Gaussian distributed over longer observation periods.

Another useful metric is the maximum time interval error (MTIE). Figure 5.12 compares the MTIE[2] of the M12+ PPS output to the phase-locked UCT GPSDO

---

[1]The noise types were unambiguously identified through an MDEV analysis.

[2]Special thanks to S. Lewis for providing his MTIE algorithm which was used to compute these results.

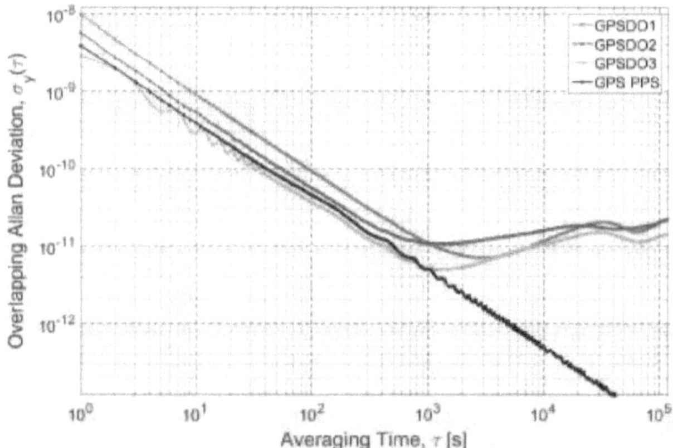

Figure 5.8: Overlapping Allan Deviation of the Free-Running UCT GPSDOs versus UTC. The short-term OCXO frequency stability is masked by the high-frequency noise of GPS PPS signal. However, at around 1000 s the OCXOs start to drift off while the GPS PPS signal keeps improving for increasing averaging times. Hence, the optimal GPSDO PLL time constant appears to be near 1000 seconds.

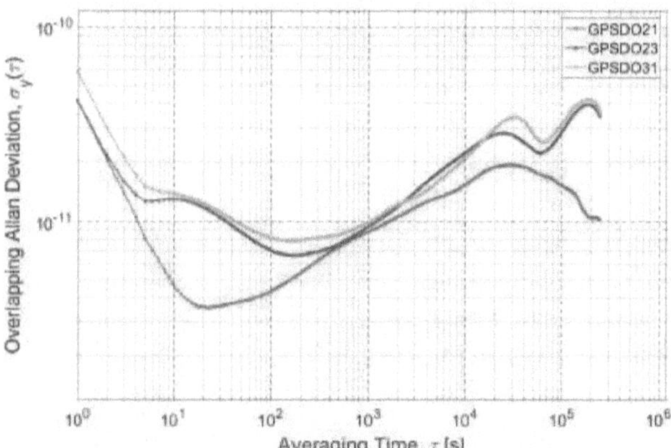

Figure 5.9: Relative Overlapping Allan Deviation of the Free-Running UCT GPSDOs.The plots of GPSDO31 and GPSDO32 show a hump between 4 seconds and 300 seconds caused by excessive Flicker FM. This hump suggests that there may be some instability in the OCXO steering circuity of GPSDO3.

for various PLL time constants over an eight hour period. The results are as expected. The MTIE of the raw GPS PPS signal remains fairly constant at about 70 ns while the sawtooth corrected PPS signal hovers between 10 ns and 20 ns. The free-running GPSDO MTIE increases on an almost linearly trajectory and crosses through the raw PPS and sawtooth corrected PPS time errors after 40 seconds and 30,000 seconds, respectively. During phase-lock, the MTIE improves with a decreasing PLL bandwidth up until the optimal time constant is reached. The MTIE for both $\tau_{PLL} = 150$ s and $\tau_{PLL} = 400$ s converge to about 10 ns. At the optimal $\tau_{PLL} = 1000$ s, the MTIE converges to just below 6 ns.

In summary, the GPSDO phase will slowly drift by a few nanoseconds around mean offset. The rate of change of the GPSDO phase decreases with decreasing PLL bandwidths. For $\tau_{PLL} = 150$ s, the time accuracy is ± 5.5 ns which equates to a bistatic baseline error of ± 1.65 m or a carrier phase variation of $\pm 4800°$ at 2.4 GHz. For $\tau_{PLL} = 1000$ s, the time accuracy improves to ± 3 ns which equates to a bistatic baseline error of ± 0.9 m or a carrier phase variation of about $\pm 2600°$ at 2.4 GHz. Hence, at $\tau_{PLL} = 1000$ s, and under ideal circumstances and with the biases nulled, it appears possible to synchronise NetRAD to a range accuracy of roughly half its range resolution across short baselines.

5.4.2 Frequency Error

This section presents the phase-locked fractional frequency performance when two UCT GPSDOs are compared. Figure 5.13 performs an eight hour fractional frequency comparison of GPSDO31 at $\tau_{PLL} = 150$ s and $\tau_{PLL} = 1000$ s. Both cases reached atomic levels in the long term with initial frequency offsets of a few parts in 10^{-13}, and frequency drifts of a few parts in 10^{-14}. Refer to Table 5.2. However, as in the free-running case, the bistatic performance will be limited by the short-term frequency accuracy of $\pm 2.6 \times 10^{-10}$ for $\tau_{PLL} = 150$ s which reduces to $\pm 1.3 \times 10^{-10}$ at $\tau_{PLL} = 1000$ s. Thus, the GPSDOs have sufficient performance when optimally tuned to meet the NetRAD Doppler requirements. However, the maximum phase gradients at $\tau_{PLL} = 150$ s and $\tau_{PLL} = 1000$ s are 224.6 °/s and 112.32 °/s, respectively. Thus, the expected maximum coherent integration loss for $\tau_{PLL} = 150$ s and $\tau_{PLL} = 1000$ s is 6.5 dB and 1.4 dB, respectively. The frequency noise is about an order of magnitude higher than expected and it should not be difficult for these low-cost GPSDOs to meet the sub 1 dB coherent integration loss requirement by

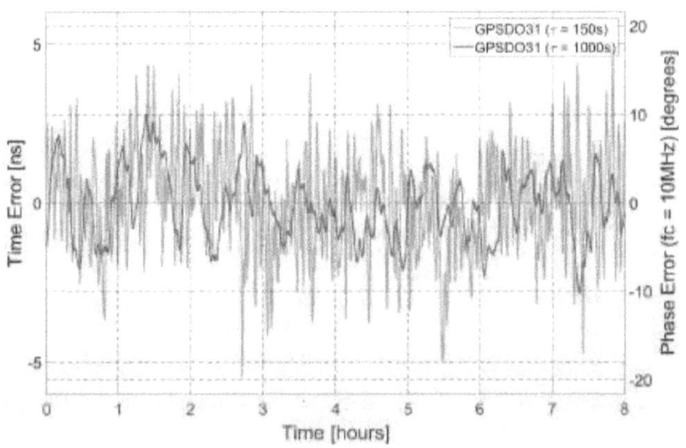

Figure 5.10: Time (Phase) Error for Phase-Locked GPSDO31 (8 hours). The rate of change of the GPSDO phase decreases with decreasing PLL bandwidths. The mean time offsets were removed.

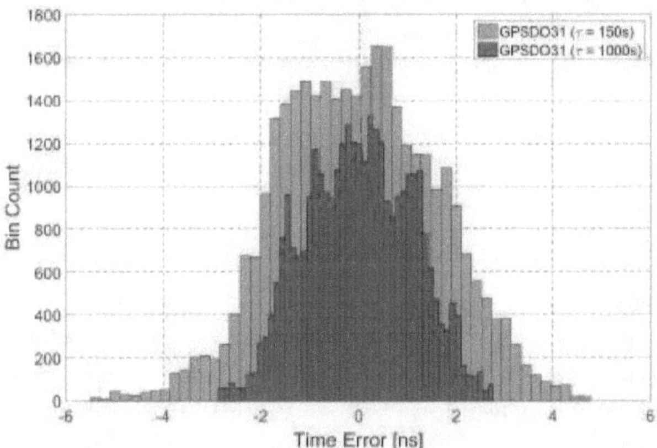

Figure 5.11: Histogram of the Time (Phase) Error for Phase-Locked GPSDO31 (8 hours). The mean time offsets were removed.

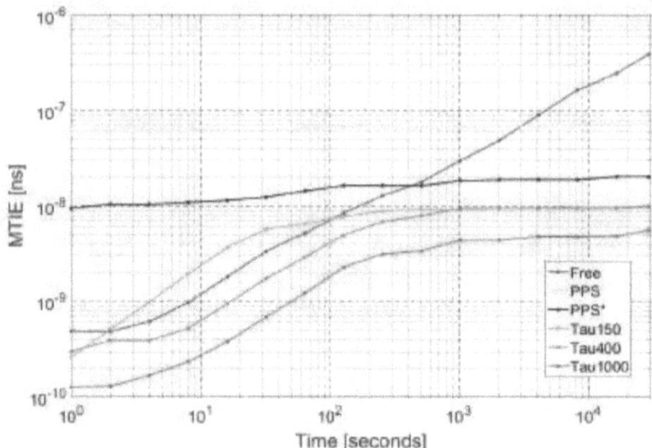

Figure 5.12: Maximum Time Interval Error for Phase-Locked GPSDO31 (8 hours).

Table 5.2: Fractional Frequency Statistics for the Phase-Locked UCT GPSDO32

	Offset	Drift (D)	Pk-Pk
$\tau_{PLL} = 150\,\mathrm{s}$	-1.78×10^{-13}	2.42×10^{-14}	$\approx \pm 2.6 \times 10^{-10}$
$\tau_{PLL} = 1000\,\mathrm{s}$	1.08×10^{-13}	1.71×10^{-14}	$\approx \pm 1.3 \times 10^{-10}$

replacing the noisy OCXO buffer amplifiers.

5.4.3 Frequency Stability

This section presents the phase-locked frequency stability performance of two UCT GPSDOs compared. During phase-lock, the GPSDO steers the internal OCXO towards the long-term trend of the GPS PPS time reference. Moreover, the GPSDO output adopts the frequency stability of the GPS PPS time reference for frequencies higher than the PLL bandwidth while retaining the initial OCXO stability within the PLL bandwidth [79]. It is essential to optimally tune the PLL bandwidth to maximise the overall frequency stability.

The MDEV performance of GPSDO21 at various PLL time constants is plotted in Figure 5.14. The MDEV plots for the sawtooth corrected PPS time references, and the free-running GPSDO21 are shown for comparison. It can be seen that the

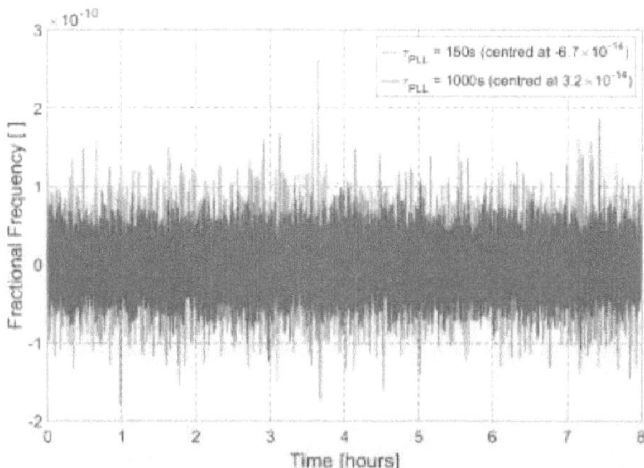

Figure 5.13: Fractional Frequency for the Phase-Locked UCT GPSDO32 (8 hours). At both PLL time constants, the initial frequency offset is a few parts in 10^{-13}, and the frequency drift is a few parts in 10^{-14}.

MDEV performance of GPSDO21 improves for increasing values of τ_{PLL} up until the optimal $\tau_{PLL} = 1000$ s is reached. For clarity, it is not shown in this plot, but it was verified experimentally that the MDEV performance rapidly deteriorates for $\tau_{PLL} > 1000$ s. The GPSDO21 achieved $\mathrm{Mod}_{\sigma_y}(\tau = 150) \leq 8 \times 10^{-12}$ and $\mathrm{Mod}_{\sigma_y}(\tau = 1000) \leq 2 \times 10^{-12}$. Better OCXO buffers should improve $\mathrm{Mod}_{\sigma_y}(\tau = 1)$ by an order of magnitude. The TDEV plot in Figure 5.15 is a useful metric derived from the MDEV plot. Here, the free-running GPSDO21 reaches a time instability of nearly 500 ns after a day of averaging. However, when phase-locked, the time instability is kept below 1 ns.

The above measurements confirm that the UCT GPSDO PLL is properly functioning and optimally tuned at $\tau_{PLL} = 1000$ s while it is tracking the GPS PPS stability well for $\tau_{PLL} \geq 1000$ s.

5.5 Phase Noise

This section considers the phase noise performance of the UCT GPSDO. The UCT GPSDO is based on a 10 MHz Oscilloquartz 8788 OCXO. For interfacing to NetRAD, the 10 MHz output is ten times multiplied to 100 MHz. This analogue multiplication

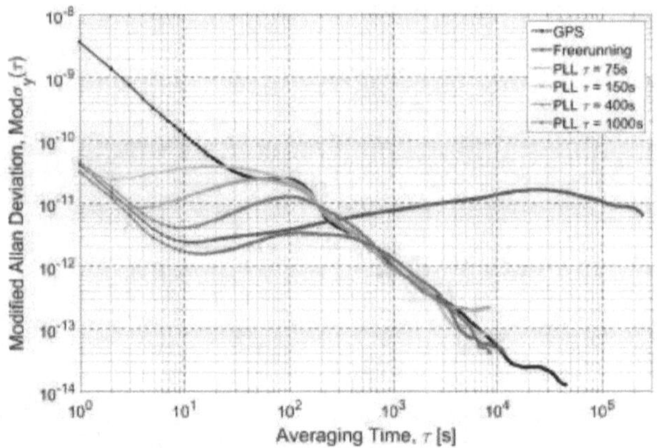

Figure 5.14: Modified Allan Deviation for Phase-Locked GPSDO21. The MDEV performance of GPSDO21 improves for increasing values of τ_{PLL} up until the optimal $\tau_{PLL} = 1000\,$s is reached.

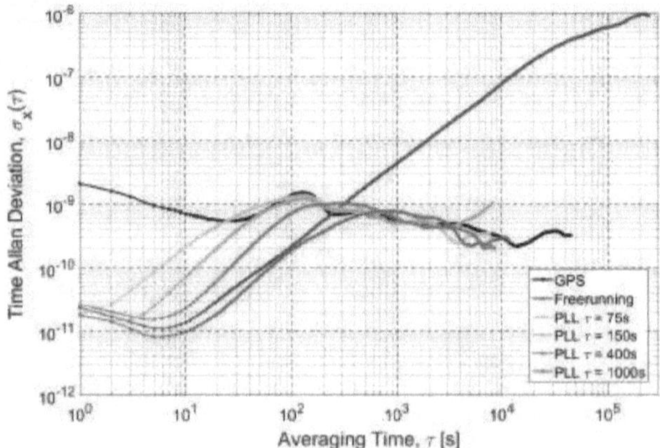

Figure 5.15: Allan Time Deviation for Phase-Locked GPSDO21. The free-running GPSDO21 reaches a time instability of nearly $500\,$ns after a day of averaging. However, when phase-locked, the time instability is kept below $1\,$ns.

increases the phase noise by 20 dBc/Hz. A Mini-Circuits ERA-6SM+ MMIC RF amplifier is used as clock buffer and an AD9512 clock distribution integrated circuit (IC) distributes the clock in a digital LVPECL format. Refer to Section 4.1.4 for a description of the UCT GPSDO clock distribution circuitry. Also, find the factory measured phase noise figures for the Oscilloquartz 8788 OCXO in Appendix B.

The 100 MHz LVPECL clock serves as frequency reference to each NetRAD node. Each node PLL synthesise its 2.4 GHz LO frequency from this reference. The 2.4 GHz PLL has a bandwidth of about 2 kHz [12]. Therefore, the close-in phase noise specification below 2 kHz of the reference clock is critically important. The phase noise of the reference clock becomes less relevant with increasing frequencies above the PLL bandwidth [79].

The phase noise was measured using an RS FSQ40 signal analyser[1] and is displayed in Figure 5.16. At 1 Hz, the level is approximately at -70 dBc/Hz and consists of Flicker FM up to about 100 Hz when White PM and Flicker PM starts to dominate. Unfortunately, there was never an opportunity to measure the phase noise of GPSDO1. However, the three GPSDOs are identical, and it is fair to assume similar performance. The bistatic phase measurements in Chapter 6 supports this assumption.

Table 5.3 compares the phase noise performance of the Oscilloqaurtz 8788 (before and after multiplication) to the final LVPECL clock output. Due to the poor performing ERA-6SM+ amplifier, the noise level of the LVPECL output is about 15 dBc/Hz to 20 dBc/Hz higher than that of the multiplier output. Nonetheless, the phase noise performance is sufficient for NetRAD measurements in Chapter 6. However, it is strongly recommend that the discrete common-base BJT amplifier is populated for future measurements.

5.6 Conclusion

This chapter presented the synchronisation performance of the UCT GPSDO. The relative time (phase), frequency, frequency stability and phase noise performance is measured at a zero baseline. The baseline GPSDO performance is recorded, and their correct operation is verified.

---

[1]Special thanks to Dr M. Ritchie and Dr F. Fioranelli for measuring the UCT GPSDO phase noise at the UCL lab in London, United Kingdom.

Table 5.3: UCT GPSDO Phase Noise

Frequency [Hz]	OCXO 10 MHz [dBc/Hz]	10× Multiplied 100 MHz [dBc/Hz]	LVPECL Output 100 MHz [dBc/Hz]
1	-100	-80	-70
10	-130	-110	-90
100	-152	-132	-115
1,000	-160	-140	-120
10,000	-165	-145	-125
100,000	-165	-145	-130

The testbench was set up carefully to minimise time biases and to ensure good GPSDO and antenna temperature tracking.

The performance of the M12+ GPS receivers is consistent with what is reported in the literature for multi-channel one-way GPS time transfer. The combined GPS antenna-receiver fixtures have time biases ranging between -2.7 ns and 3.1 ns. The part-to-part skew within the GPSDO EP circuitry contributes an additional 4 ns. However, these time biases are constant and can be removed through calibration. The raw GPS PPS signals have a time accuracy of about 80 ns translating to a bistatic baseline error of roughly 24 m. Real-time sawtooth correction can improve the time accuracy to about 28 ns which translates to a bistatic baseline error of roughly 8.4 m. Moreover, sawtooth removal improves the relative peak-to-peak and the RMS timing errors by a factor of 3 and 5, respectively. Sawtooth removal improves the ODEV by a factor of five.

It was demonstrated that a GPSDO could self-calibrate[1] its optimal PLL bandwidth by comparing the free-running OCXO to UTC using the built-in phase detection circuitry. The optimal PLL time constant for the UCT GPSDO was determined to be around 1000 s.

---

[1]Rochat, Leuenberger, Stehlin [68] described an Rb-based GPSDO which auto-tunes the PLL bandwidth continuously based on the frequency stability of the input PPS reference.

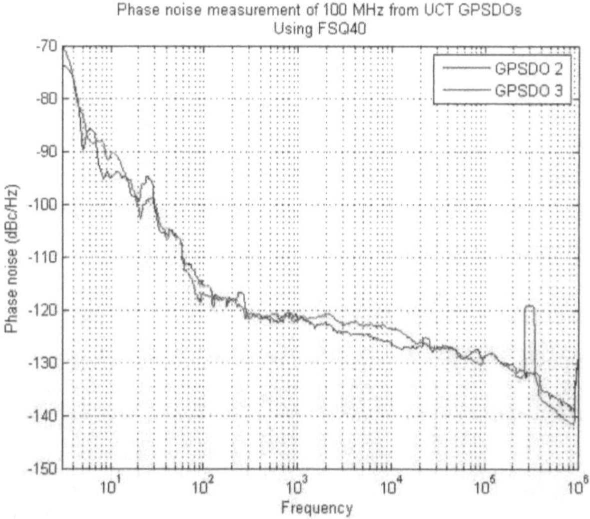

Figure 5.16: UCT GPSDO Phase Noise (100MHz LVPECL Output). It is not immediately clear why GPSDO3 has an excursion at 30 kHz but it is assumed to be a manufacturing issue.

As expected, the quartz-based GPSDOs are not suitable for bistatic time synchronisation during extended hold-over periods with the UCT GPSDOs drifting by as much as a few microseconds per day. Nonetheless, the hold-over frequency accuracy (3×10^{-10}) is sufficient to achieve a bistatic Doppler accuracy of a few Hertz at radar frequencies of a few gigahertz during a day of hold-over. The free-running frequency stability of GPSDO1 and GPSDO2 was as expected where White PM, Flicker PM and Flicker FM noise are present up until about 10 seconds after which Random Walk FM noise starts to dominate. However, an ODEV Flicker FM hump between 4 seconds and 300 seconds suggests and instability in the steering circuitry of GPSDO3. This anomaly also explains the somewhat higher hold-over time and frequency drift rates for GPSDO3.

The phase-locked UCT GPSDOs were calibrated at various PLL bandwidths. For $\tau_{PLL} = 150$ s, the peak-to-peak the time accuracy is ± 5.5 ns which equates to a bistatic baseline error of ± 1.65 m or a carrier phase variation of $\pm 4800°$ at 2.4 GHz. The rate of change of the GPSDO phase decreases with decreasing PLL bandwidths and the time accuracy improves to ± 3 ns for $\tau_{PLL} = 1000$ s. This equates to a bistatic baseline error of ± 0.9 m or a carrier phase variation of about $\pm 2600°$ at 2.4 GHz. As expected, the phase-locked UCT GPSDOs achieved atomic level long-term frequency accuracy (10^{-13}) and drift rates (10^{-14}). However, the bistatic Doppler performance will be limited by the higher short-term frequency accuracy of $\pm 2.6 \times 10^{-10}$ for $\tau_{PLL} = 150$ s which reduces to $\pm 1.3 \times 10^{-10}$ at $\tau_{PLL} = 1000$ s. The maximum phase gradients at $\tau_{PLL} = 150$ s and $\tau_{PLL} = 1000$ s are $224.6°/$s and $112.32°/$s, respectively. Thus, the expected maximum coherent integration loss for $\tau_{PLL} = 150$ s and $\tau_{PLL} = 1000$ s is 6.5 dB and 1.4 dB, respectively. A frequency stability of $\mathrm{Mod}_{\sigma_y}(\tau = 150) \leq 8 \times 10^{-12}$ and $\mathrm{Mod}_{\sigma_y}(\tau = 1000) \leq 2 \times 10^{-12}$ was measured. The phase noise performance of the 100 MHz UCT GPSDO is not stellar at -70 dBc/Hz at 1 Hz where the Flicker FM noise drops off to -115 dBc/Hz at 100 Hz. The UCT GPSDO phase noise can be improved by between 15 dBc/Hz to 20 dBc/Hz by removing the poorly selected temporary clock buffer and populating the discrete common-base BJT buffer amplifier. This superior amplifier is also expected to improve the ODEV performance for averaging times below 10 s, as well as, the frequency noise by roughly an order of magnitude. Moreover, low-cost quartz GPSDOs should easily achieve a short-term frequency accuracy of a few parts in 10^{-11} ensuring a sub 1 dB coherent SNR loss when synchronising NetRAD.

In summary, the UCT GPSDO were carefully calibrated, and it was confirmed that the critically damped PLL is optimally tuned at $\tau_{PLL} = 1000$ s. The OCXO steering circuitry of GPSDO3 performs somewhat worse than the others. Further, the short-term frequency stability, accuracy, and phase noise of all the GPSDOs are expected to improve significantly by with better OCXO buffer amplifiers. Nonetheless, low-cost quartz GPSDOs are expected to meet all the frequency and phase requirement of NetRAD for ideal circumstances across zero-baselines. However, it seems improbable that a sub-3-metre range accuracy can be achieved under field conditions. In Chapter 6, the UCT GPSDOs will be tuned to $\tau_{PLL} = 150$ s and used to synchronise NetRAD. The radar performance will be compared to the zero-range calibrations of this chapter.

Chapter 6

Synchronising NetRAD using the UCT GPSDOs

In this chapter, NetRAD is synchronised using the UCT GPSDOs. A tristatic experiment is set up in Simon's Bay, Cape Town. The achieved synchronisation is compared to the zero-baseline measurements of Chapter 5, and it is assessed how well the GPS synchronised NetRAD meets the requirements defined in Section 3.6. Moreover, the sidelobe breakthrough and strong reflections from the Roman Rock lighthouse are used to assess the achieved time, phase, and frequency synchronisation. LOS phase compensation (see Section 3.3) is applied to the bistatic data, and the phase, frequency, Doppler phase noise and Doppler performance of the monostatic, bistatic and LOS phase compensated Roman Rock reflections are compared. Finally, bistatic pulse integration in the presence of a constant frequency offset (see Section 3.4) is studied using real bistatic data.

Originally, NetRAD was a cabled system which limited the baselines to 50 metres [12, 13, 29]. In a subsequent collaboration between UCL and UCT, the cabled NetRAD was converted to an entirely wireless network radar which made baselines of up to a few kilometres possible. A long range WIFI network carries the data and control signals. Each node in the master-slave based wireless NetRAD is fitted with a UCT GPSDO providing time and carrier synchronisation [27]. Moreover, network-wide time synchronisation is established using UCT GPSDO EP [35].

The tristatic geometry set up in Simon's Bay has a maximum baseline of 2.3 km. The maximum bistatic range to the Roman Rock lighthouse is about 3.4 km. The well-aged GPSDOs were set up for one-way GPS time transfer and each node self-surveyed its coordinates. Long PLL time constants proved to be impractical due to excessive lock-in times. Subsequently, the GPSDOs were all tuned to a suboptimal $\tau_{PLL} = 150\,\text{s}$. Therefore, the zero-range GPSDO calibrations in Chapter 5 included recordings for $\tau_{PLL} = 150\,\text{s}$ to allow meaningful comparisons to this bistatic data.

Time precision could not be measured without independently surveyed site positions. However, the time accuracy was measured using the change in the breakthrough time of arrival across a static baseline. The maximum measured bistatic time drift of $\pm 5\,\text{ns}$ about the mean over a 20-minute period is consistent with the zero-baseline calibration for $\tau_{PLL} = 150\,\text{s}$.

Next, the phase of the Roman Rock reflections is considered. The bistatic phase drift is consistent with the zero-range calibrations. Moreover, the phase slopes for $\tau_{PLL} = 150\,\text{s}$ were kept below $95\,°/\text{s}$ keeping the coherent SNR loss below $1\,\text{dB}$.

The fractional frequency is computed by taking the first difference of the pulse-to-pulse phase in units of time. The bistatic frequency offset of -2.03×10^{-11} was an order of magnitude lower than required. The frequency drift was negligible.

Due to limited time and equipment, the close-in IF phase noise was never measured. Instead, the Doppler phase noise was estimated by calculating the power spectral density (PSD) of the matched filtered data and scaling it to units of dBc/Hz. This Doppler phase noise is used as a Doppler performance measure to compare the monostatic, bistatic, and LOS phase compensated bistatic cases. However, further analysis is required to confirm. However, the Doppler phase noise cannot be directly related to the IF phase noise. It is left as future work to analyse the IF phase noise.

The bistatic Doppler analysis was done using an FFT spanning the entire 130-second recording. This unrealistically long Doppler integration results in a very high Doppler resolution and also exaggerates the GPSDO phase dynamics. The bistatic Doppler noise floor is about $30\,\text{dB}$ higher than the monostatic. However, the small target Doppler performance is reasonable when not masked by larger targets.

Bistatic LOS phase compensation proved very successful across such short baselines. The compensated bistatic phase and frequency data reached near monostatic levels. Moreover, LOS phase compensation improved the high-frequency bistatic Doppler phase noise by $20\,\text{dBc/Hz}$ to reach the monostatic level.

Finally, the effect of a constant bistatic frequency offset on both non-coherent and coherent integration is studied using real bistatic data. A $1/f$ integrated noise component appears advantageous under certain conditions. The results compared well to the theory developed in Section 3.4, but more data and analysis is required.

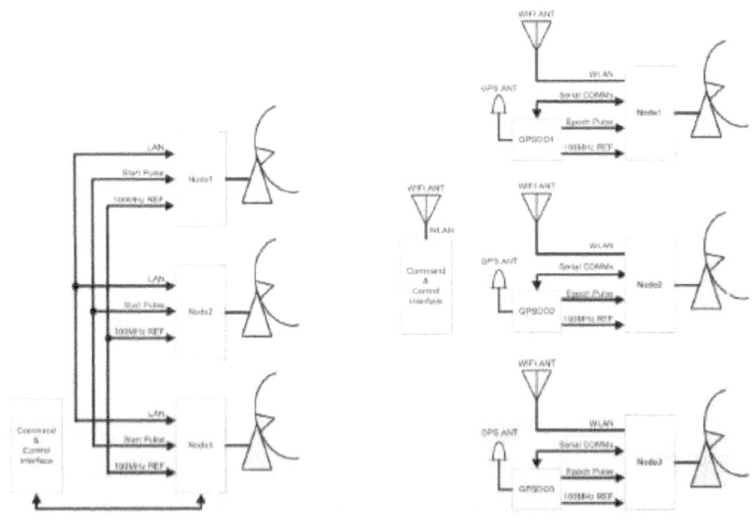

(a) The Cabled NetRAD System.   (b) The Wireless NetRAD System.

Figure 6.1: The Conversion of NetRAD to a Fully Wireless Networked Radar.

6.1 The Wireless NetRAD System

NetRAD is a coherent tri-node pulsed-Doppler network radar. The carrier frequency is at S-band (2.4 GHz) with a bandwidth of 50 MHz. NetRAD has been under development at UCL since the early 2000's and was built using low-cost COTS hardware on a budget of £4000. [12, 13, 29] In the original design, the nodes were connected via 50 m long CAT6 ethernet cables. These cables carried the radar data and control signals, frequency reference and trigger pulses achieving good carrier coherence. Each node has a basic transmit power of 200 mW, but a 500 W power amplifier was later acquired for one of the nodes. Refer to Table 3.1 for the basic NetRAD specifications. The higher powered 500 W amplifier allowed small target detections at much longer ranges of up to a few kilometres. However, at such target distances, the 50-metre baselines limited the bistatic angles to only a few degrees. [27]

In a subsequent collaboration between UCL and UCT, the cabled NetRAD system was converted to an entirely wireless network radar which made baselines of up to a few kilometres possible. Moreover, the system was changed from a peer-to-peer to a master-slave based system. Commercial 5.3 GHz (IEEE 802.11a) WIFI links were used for the data and control signals, and the UCT developed GPSDOs were used for carrier phase and time synchronisation. This upgrade to NetRAD was an ambitious task and required a team of students, academics and industry experts. The author was responsible for the GPS-synchronisation [27]. The added flexibility in radar geometry enabled several bistatic sea clutter and small sea vessel trials along the Southern coasts of England and South Africa [100–105].

In the cabled system, the control interface was built-in to one of the nodes. Refer to Figure 6.1a. This node controls all the other nodes and distributes the 100 MHz frequency reference. Each node PLL-synthesises its local 2.4 GHz carrier, PRF and sampling frequencies from this 100 MHz reference. After receiving the radar configuration, each node would then wait for the control node to produce the 'start pulse'. This 'start pulse' edge signals the start of an experiment at which point each node starts to sample for the set amount of samples and pulses. [12, 27]

The wireless system uses a master-slave topology. The master command and control unit is separate while the slave nodes all have identical hardware and software. Refer to Figure 6.1b. A long-range WIFI network carries data and control communications between the master control unit and slave nodes. This network enables NetRAD to be controlled and operated remotely. Each node is fitted with a UCT GPSDO providing a GPS synchronised 100 MHz frequency reference. A serial interface gives the master unit real-time access to each GPSDO. Moreover, the GPSDO PLL can be monitored and adjusted in real-time, and the EP can be remotely armed for network-wide time synchronisation. Before each experiment, the radar parameters and the future experiment start time are uploaded to each node. The GPSDO then generates an EP edge at the specified start time, and each node would start to sample for the set amount of samples and pulses. [27] Refer to Section 4.2.4 for a description of the EP mechanism [35].

6.2 Tristatic Radar Geometry

A tristatic NetRAD experiment was set up in Simon's Bay just outside Cape Town, South Africa. The purpose of this experiment was to assess the performance of the GPS synchronised wireless NetRAD. See Figure 6.2 for a Google Earth view of this tristatic radar geometry with an estimate of the overlapping beam areas.

The monostatic node (Node3) with the 500 W transmitter was set up on the

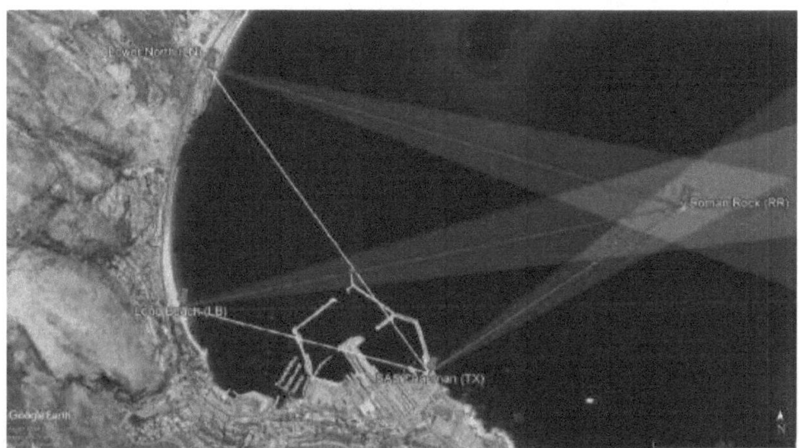

Figure 6.2: Tristatic Radar Geometry, Simon's Bay. The approximate baselines towards Long Beach and Lower North were 1.7 km and 2.3 km, respectively. All the nodes had identical antennas with their 10° beams centred on the Roman Rock (RR) lighthouse. The monostatic node had an unobstructed LOS to the passive nodes.

roof of the SAS Chapman building (TX) in the Simon's Town naval base. The two passive nodes (Node1 and Node2) were placed at Long Beach (LB) and the Lower North Battery (LN), respectively. The approximate baselines towards Long Beach and Lower North were 1.7 km and 2.3 km, respectively. All the nodes had identical antennas with their 10° beams centred on the Roman Rock (RR) lighthouse. Vernier antenna turntables were used to measure the true North antenna bearings with an accuracy of less than one degree.

The GPSDOs were configured for one-way GPS time transfer. The node co-ordinates were self-surveyed. These self-surveyed estimates are the best available positional information together with maps and antenna bearings. However, due to on-the-day time constraints, the GPSDOs could not be left to complete the entire 3-hour self-survey. After self-surveying the antenna coordinates for between 30 and 60 minutes, each GPSDO was manually put in 'position hold' mode. Nonetheless, the coordinates converged after about 15 minutes of averaging. It is believed to be a sensible trade-off between the survey time and positional accuracy. A software problem resulted in the GPS data for Node1 being lost. However, the self-surveyed GPS data is of somewhat limited use without independently surveyed control po-

Table 6.1: Tristatic Radar Geometry (GPS)

	Lat	Lon	Alt[a]	Baseline	Range[b]	Bearing	β
LB[c]	-	-	-	-	-	-	-
LN	-34.17376°	18.42793°	17.48 m	2337 m	2626 m	105.66°	49.41°
TX	-34.1903202°	18.44305083°	28.81 m	-	1888 m	56.25°	-

[a]The GPS altitude is converted to the orthometric height using Simon's Bay geoidal height of 30.947 m. However, GPS measured altitude is known to be inaccurate.

[b]Mono- and bistatic ranges to Roman Rock. Roman Rock's coordinates were estimated using Google Earth as -34.1812194° and 18.46011667°.

[c]The Node1 GPS data did not record during this experiment.

sitions and the data loss did not hinder the experiment. The GPS data is mainly convenient for comparative purposes. The radar geometry based on the GPS data is summarised in Table 6.1. In this table, Roman Rock's coordinates were estimated using Google Earth.

The GPSDOs were well-aged for about a month. GPSDO1 was paired with Node1 (LB), GPSDO2 was paired with Node2 (LN), and GPSDO3 was paired with Node3 (TX). Throughout this chapter, the convention is followed where GPSDO32 is the relative error GPSDO32 = GPSDO3 − GPSDO2. Long PLL time constants proved to be impractical due to excessive lock-in times. Subsequently, the GPSDOs were all tuned to a suboptimal $\tau_{PLL} = 150$ s. The zero-range GPSDO calibrations in Chapter 5 included recordings for $\tau_{PLL} = 150$ s to allow meaningful comparisons to this bistatic data. Later, a quick-locking PLL (see Section 4.3.4) was developed enabling the practical use of $\tau_{PLL} \geq 1000$ s [35]. This algorithm is now successfully implemented in the newer NeXtRAD system [33]. In all cases, the GPSDO EP function (see Section 4.2.4) provides network-wide time synchronisation [35].

The monostatic node had an unobstructed LOS to the passive nodes. There was a strong sidelobe breakthrough along each baseline. This breakthrough along with the Roman Rock reflections was used to test the bistatic synchronisation and LOS phase compensation.

6.3   Bistatic Synchronisation Performance

This section presents the NetRAD performance when synchronised using the UCT GPSDOs. It compares the achieved synchronisation to the zero-baseline measurements of Chapter 5, and assess if and how well the GPS synchronised NetRAD system meets the performance requirements defined in Section 3.6.

The sidelobe breakthrough and Roman Rock reflections are used to assess the achieved time, phase and frequency synchronisation. After assessing the time synchronisation, LOS phase compensation (see Section 3.3) is applied to the Roman Rock returns. The phase, frequency, Doppler phase noise and Doppler performance of the monostatic, bistatic and LOS phase compensated Roman Rock reflections were compared. Finally, bistatic pulse integration in the presence of a constant frequency offset (see Section 3.4) is studied using real data.

The Simon's Bay geometry (see Section 6.2) is used unless stated otherwise. These tests were conducted concurrently with unrelated Doppler trials of small sea vessels which limited the PRF to 1 kHz and the maximum possible recording time to 130 seconds. Refer to Table 3.1 for the basic NetRAD parameters.

6.3.1   Time Error

This section discusses the time synchronisation performance of GPSDO synchronised NetRAD. A comparison of the radar- and GPS-measured radar geometries gives a rough estimate of the time precision. Refer to Table 6.1 and Table 6.2. The time accuracy is measured using the change in the breakthrough time of arrival across a static baseline.

A 130-second range-time intensity (RTI) plot for each node in the Simon's Bay geometry is given in Figure 6.3. In the monostatic RTI plot, the compressed pulse transmission peaked 1.07 µs after the sampling started. The measured monostatic range to Roman Rock is 1899 m. Assuming a network-wide GPS time synchronisation better than one range bin ($x_o \leq 10\,\text{ns}$), the estimated transmission time at bin 107 is indicated by the blue lines (TX) in the passive RTI plots. In both passive RTI plots, the breakthrough appears as the first target. Roman Rock is the second target. In each case, the TX-BR time delay represents the bistatic baseline, and the BR-RR time delay represents the total bistatic time delay. The radar measured geometry is summarised in Table 6.2.

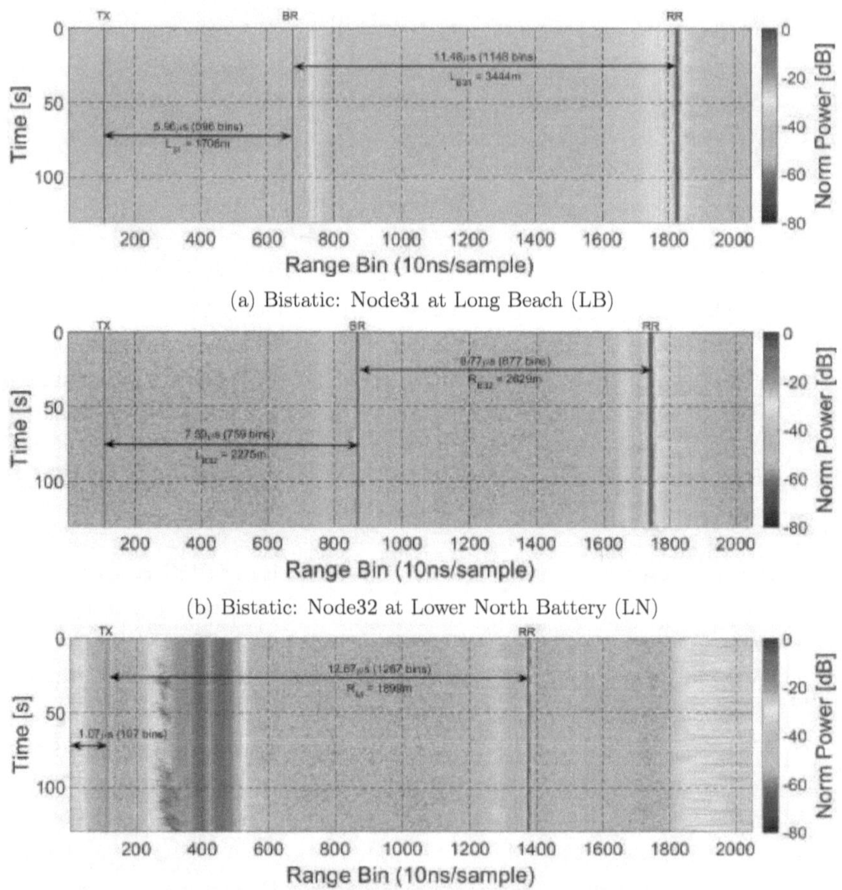

(a) Bistatic: Node31 at Long Beach (LB)

(b) Bistatic: Node32 at Lower North Battery (LN)

(c) Monostatic: Node33 at SAS Chapman (TX)

Figure 6.3: Tristatic Range-Time Intensity Plots for the Simon's Bay Geometry.

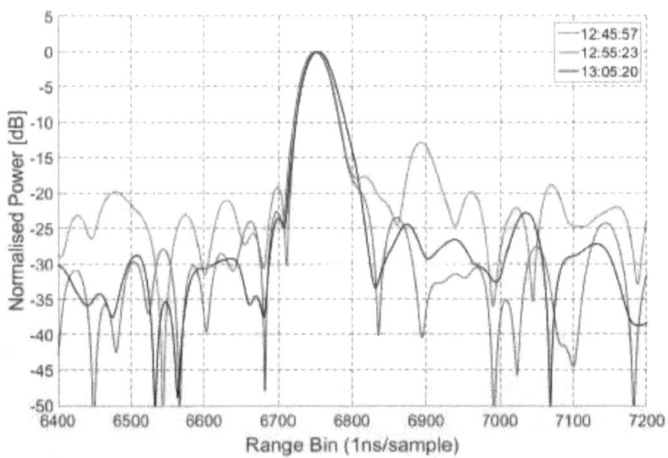

(a) Bistatic: Node31 at Long Beach (LB). The peak breakthrough amplitute varied by ±2 ns over the 20-minute period. The measurement at 12:58:56 did not record because a train passed through the beam.

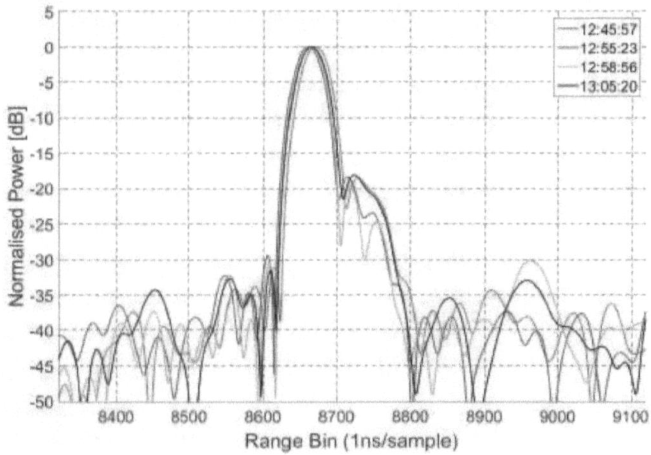

(b) Bistatic: Node32 at Lower North Battery (LN). The peak breakthrough amplitute varied by ±5 ns over the 20-minute period.

Figure 6.4: Overlapping Range Profiles of the Breakthrough for Various Experiments. These plots are zero-pad interpolated to reduce the range bin size to 1 ns.

Table 6.2: Tristatic Radar Geometry (radar)

	Baseline	Range	Bearing	β
LB (Node1)	1706 m	3444 m	78.5°	21.94°
LN (Node2)	2275 m	2629 m	106.5°	49.46°
TX (Node3)	-	1899 m	57°	-

One cannot measure bistatic time precision without independently known site positions. However, in Chapter 5 a sub-10 ns mean GPSDO time offset was measured comprising of the GPS time precision ($\leq$ 6 ns) plus the EP skew ($\approx$ 4 ns).[1] The 62 m (206 ns) difference between the radar- and GPS-measured TX-LN baselines is therefore noteworthy. Contrarily, there is only a 3 m (10 ns) difference in the TX-RR-LN bistatic range. Also, the radar- and GPS-measured antenna bearings agree to within 1°. It suggests that the GPS position is accurate to within a few meters and that the sampling origins are offset by 206 ns. This does not agree with the measured sub-10 ns zero-baseline time precision or with the results reported in the literature [19, 60]. Further experiments with precisely known site locations are necessary to confirm.

The baselines are constant, and a change in the breakthrough arrival represents the time accuracy.[2] Moreover, the breakthrough and Roman Rock reflections are kept within a single range bin during a 130-second recording. This result was consistent across multiple experiments. The experiment-to-experiment propagation delays are considered to assess the longer-term time drift. Moreover, Figure 6.4 plots the overlapping range profiles for the breakthrough of four independently started experiments over a 20-minute period. The peak breakthrough amplitude for Node1 and Node2 varied by $\pm$2 ns and $\pm$5 ns, respectively. The above measured time accuracies are consistent with the $\pm$5.5 ns zero-range GPSDO time drift about the mean for $\tau_{PLL} = 150$ s.

In summary, the achieved time precision requires further investigation while the time drift about the mean is as expected. As expected, it is not possible to achieve

---

[1] GPSDO PLL calibration can null these static equipment offsets. However, the mean offsets were not nulled during these trials.

[2] It is assumed here that the short baseline propagation and multipath errors are much smaller than the GPSDO time error.

a sub-10 ns time accuracy at $\tau_{PLL} = 150\,\text{s}$ with self-surveyed coordinates. Short baseline sub-10 ns time synchronisation may be possible at $\tau_{PLL} = 1000\,\text{s}$, but it will require professionally surveyed coordinates [31] and temperature stabilised GPS antennas [61]. For accurately known baselines of a few kilometres, it is probably more practical to use the breakthrough to achieve sub-10 ns time synchronisation.

6.3.2 Range Error due to Time Offset

Figure 6.5 demonstrates the relationship between a Tx-Rx time offset and bistatic range. This RTI plot is of a sea clutter measurement using a different geometry with a baseline of a few hundred metres. The simultaneously recorded Tx-Rx GPSDO PLL time error is superimposed onto the RTI image. Initially, both GPSDOs are phase-locked with the direct breakthrough time delay constant at $\tau_{B32} = 2.1\,\mu\text{s}$. At 17 seconds, GPSDO2 was offset by 411 ns (123 m) and left to re-acquire phase lock. As expected, the RTI plot is distorted proportionally to the GPSDO32 PLL time error. Hence, the relative GPSDO PLL time error is a good measure of the range error.

6.3.3 Range Creep due to Frequency Offset

In Section 3.2, it is discussed how a bistatic frequency offset mimics a relative Tx-Rx velocity. The bistatic RTI plot in Figure 6.6 demonstrates this effect. This measurement used a different geometry with a baseline of a few hundred metres. At the start of this experiment, both GPSDOs were phase-locked resulting in the static breakthrough visible at $\tau_{B31} = 1.44\,\mu\text{s}$. The simultaneously recorded GPSDO31 PLL error is unwrapped and superimposed onto the RTI image. At 56 seconds, the transmitter's GPSDO frequency is offset by 4.269×10^{-7} (about 4 Hz at 10 MHz). This frequency offset induced a false Tx-Rx velocity of 128 m/s.

One can make some observations from Figure 6.6. Firstly, the transmitter's reference frequency is higher than that of the receiver. Consequently, the transmitting and receiving PRFs are offset and wrapping such that there is an apparent increase in the bistatic baseline. Initially, the maximum bistatic range was 5712 m. However, after introducing the frequency offset the maximum range decreased linearly with time and at 100 seconds the maximum range is reduced to a mere 84 m. Secondly, the breakthrough and target reflections are equally distorted. Thus, the bistatic

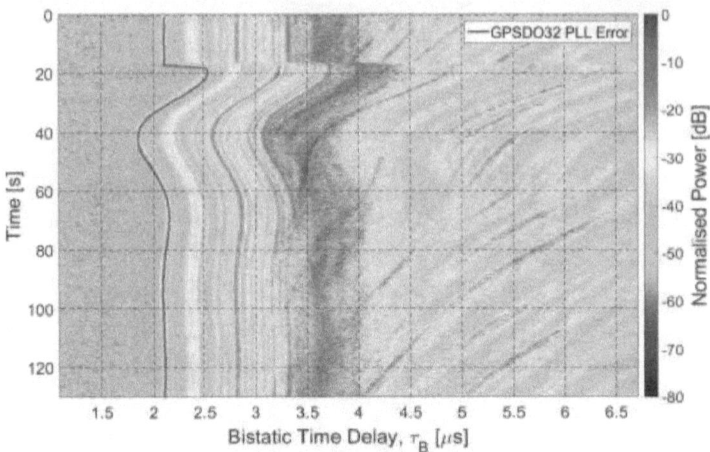

Figure 6.5: A GPSDO Locking Transient Proportionally Distorts the Bistatic Time Delay. The relative GPSDO PLL time error is a good measure of the range error. The GPSDO32 PLL time error lags by one second because the GPSDO only updates once every second.

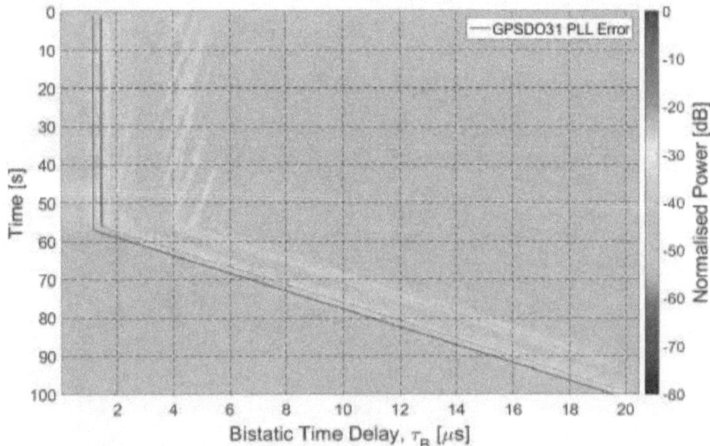

Figure 6.6: A Bistatic Frequency Offset Induces a False Tx-Rx Velocity. At 56 seconds, the transmitter's GPSDO frequency is offset by 4.269×10^{-7} (about 4 Hz at 10 MHz). This frequency offset induced a false Tx-Rx velocity of 128 m/s. The superimposed PLL error is offset for clarity.

range is unaffected given that the breakthrough arrival time is known for every pulse.

6.3.4 Phase Error

This section considers the carrier phase performance of the GPSDO synchronised NetRAD. Both the mono- and bistatic Roman Rock returns are compared.

The unwrapped Roman Rock phase is plotted in Figure 6.7. As expected, the GPSDO phase steering causes the bistatic phase to drift during the 130-second recording. The maximum phase excursion for the three plots ranged between 2506°, 3124° and 5689° at the 2.4 GHz carrier. These angles correspond to 10.44°, 13.02° and 23.7° at 10 MHz. A phase excursion of 23.7° at 10 MHz relates to time excursion of 6.58 ns. These values correspond well with the zero-baseline phase accuracy of ±20° at 10 MHz recorded in Section 5.4.1 over an eight hour period. The phase discontinuities at around 20 seconds and 125 seconds in the plot of Node31 are assumed to be phase reversals caused by cable movement.

Consistent with the assumption in Section 3.6.4, it appears valid to assume linear phase slopes over integration periods of up to a few seconds. The maximum linear phase slope is -92 °/s and appears between 10 seconds and 50 seconds for plot Node32 at 13:05:20. This phase slope translates to a frequency offset of 1.1×10^{-10}, or a Doppler error of 0.256 Hz at 2.4 GHz, or a radial target velocity error of 0.115 km/h, and the integrated coherent SNR loss should be below 1 dB. This frequency offset is below the zero-range measured maximum short-term frequency accuracy of $\pm 2.6 \times 10^{-10}$. However, note that the relative GPSDO frequency slowly drifts within this range over minutes and the measured Doppler phase slope may fall anywhere within this range. Thus, the results are consistent with the zero-range measurements. However, much longer phase recordings are necessary to confirm. Adjusting the GPSDOs to $\tau_{PLL} = 1000$ s is expected to reduce the rate of phase change improving the frequency accuracy.

Figure 6.8 compares the bistatic and monostatic carrier phases for a 5-second Roman Rock recording. As expected, the monostatic carrier phase exhibits significantly less noise and drift when compared to the bistatic. The monostatic carrier phase has a peak-to-peak variation of approximately 15° with no apparent drift whereas the bistatic carrier phase drifts by approximately -14 °/s. However, it is expected that multipath across the moving ocean, as well as wave motion at Ro-

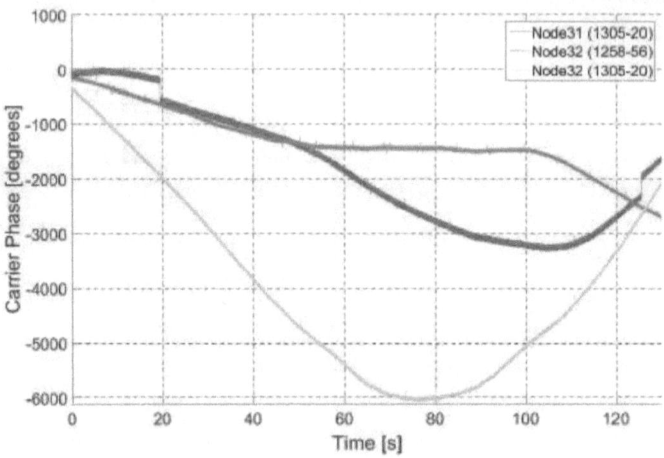

Figure 6.7: Bistatic Carrier Phase for Roman Rock. The maximum phase excursion of 23.7° at 10 MHz and correspond well with the zero-baseline GPSDO phase accuracy of $\pm 20°$ at 10 MHz for $\tau_{PLL} = 150$ s. The phase discontinuities at around 20 seconds and 125 seconds in the plot of Node31 are assumed to be phase reversals caused by cable movement.

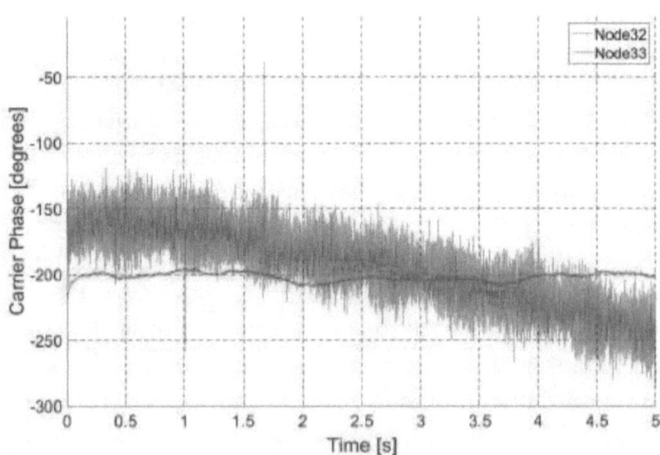

Figure 6.8: Carrier Phase for Roman Rock: Bistatic versus Monostatic. The monostatic phase exhibits significantly less noise and drift when compared to the bistatic.

man Rock, are at least partially responsible for the monostatic phase variation. The short-term ($<$0.1 s) peak-to-peak noise is 2.5° and 90° for the monostatic and bistatic, respectively.

In summary, the phase performance is consistent with the zero-baseline measurements. In these limited recordings, the GPSDOs set at $\tau_{PLL} = 150$ s, kept the phase slopes below 95 °/s to achieve the sub-1 dB integrated, coherent SNR loss requirement set in Section 3.6.4. However, from the zero-range calibrations, one can expect frequency errors as large as 2.6×10^{-10} resulting in a bistatic coherent SNR loss of up to 6.5 dB. Also, as expected, the bistatic Doppler phase noise is significantly higher than that of the monostatic.

6.3.5 Line-of-Sight Phase Compensation

A LOS phase compensation technique where the breakthrough along a static baseline is used to remove the GPSDO induced target phase dynamics is described in Section 3.3. This section applies LOS phase compensation to the GPSDO synchronised NetRAD data. Moreover, the breakthrough and bistatic Roman Rock reflections were captured within a single pulse, and then both the breakthrough and Roman Rock phases are compared before and after LOS compensation.

Figure 6.9 shows both the breakthrough and Roman Rock phases before and after LOS phase compensation.[1] The uncompensated breakthrough and Roman Rock phase returns are nearly identical but offset by roughly 110°. LOS phase nulling removes the slow varying GPSDO phase dynamic resulting in a bistatic phase with similar performance to that of the monostatic. The compensated bistatic phase has a peak-to-peak phase variation of 27.4° over 130 seconds and short-term ($<$0.1 s) peak-to-peak variation of 3.6°. The medium-term phase variation in the LOS compensated bistatic case is larger than that of the monostatic Roman Rock returns. However, this is expected because LOS phase compensation is additionally subject to multipath and propagation effects across the Tx-Rx baseline. Hence, the efficacy of LOS phase compensation is expected to deteriorate with increasing baselines. Further, it is not shown here, but similar results are achieved when using Roman Rock as the phase reference for phase compensation of the breakthrough. Thus, suggesting that a large and static target is a suitable phase reference when

---

[1]In this chapter, LOS phase compensation will be denoted by an asterisk in all figures and tables.

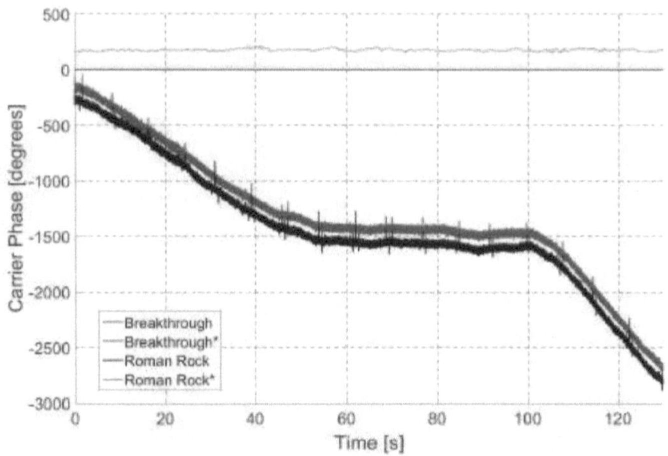

(a) Bistatic Carrier Phase (130 s recording).

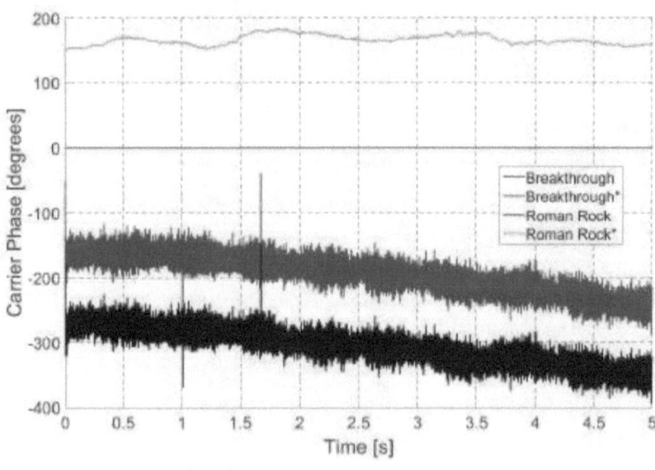

(b) Bistatic Carrier Phase (5 s recording).

Figure 6.9: Line-of-Sight Carrier Phase Compensation. The breakthrough along a static baseline is used to remove the GPSDO induced target phase dynamics. The breakthrough and Roman Rock phases before and after compensation are compared.

Table 6.3: Fractional Carrier Frequency Statistics

	Offset	Drift (D)	Pk-Pk	Spurs
Monostatic	-4.77×10^{-12}	5.69×10^{-14}	$\pm 2 \times 10^{-9}$	$\pm 1.8 \times 10^{-8}$
Bistatic	-2.03×10^{-11}	-4.96×10^{-14}	$\pm 4.8 \times 10^{-8}$	$\pm 1.8 \times 10^{-7}$
Bistatic*	1.39×10^{-12}	-1.76×10^{-14}	$\pm 2 \times 10^{-9}$	-

there is no direct LOS visibility. Phase correction does not require exact knowledge of the target location, only that the target is static with strong returns.

Thus, LOS phase compensation appears to be successful when used in combination with low-cost quartz GPSDOs to synchronise bistatic systems. Moreover, the bistatic phase performance improves to near monostatic levels. The following sections investigate this further by comparing the frequency, Doppler phase noise and Doppler performance of the monostatic, bistatic, and LOS phase compensated cases.

6.3.6 Frequency Error

This section considers the Tx-Rx carrier frequency of the GPSDO synchronised NetRAD. The fractional frequency is computed by taking the first difference of the pulse-to-pulse phase in units of time. Figure 6.10 is a 130-second fractional frequency comparison of the monostatic, bistatic and LOS phase compensated bistatic returns of Roman Rock. The data statistics are summarised in Table 6.3. The offset and drift were determined using a straight line fit.

The initial monostatic frequency offset and peak-to-peak frequency noise are roughly an order of magnitude higher than the measured zero-baseline GPSDO readings. Also, the monostatic returns contain some frequency spikes that were absent in the zero-baseline measurements. These added offsets, noise and spurs likely originate from the 2.4 GHz PLL-synthesiser internal to each node. Moreover, the frequency offset is likely due to the finite frequency resolution of the N/M synthesiser, and highlights the importance of designing the radar subsystems to match that of the frequency reference. However, it is reasonable to expect the high-frequency frequency noise on the carrier to always be higher than that of the STALO.

The initial bistatic frequency offset for this measurement is -2.03×10^{-11} and is about four times higher than the monostatic, but it remains negligible at approx-

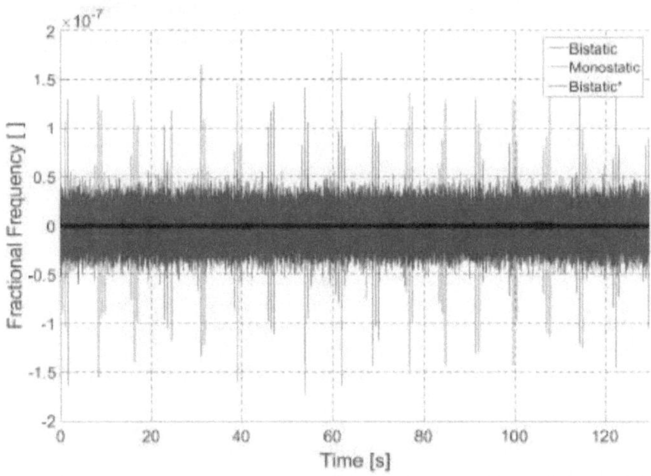

Figure 6.10: Fractional Carrier Frequency. The monostatic, bistatic, LOS phase compensated bistatic carrier frequencies are compared. LOS phase compensation restores the bistatic fractional frequency to monostatic levels.

imately $0.048\,\text{Hz}$ at $2.4\,\text{GHz}$. However, from the zero-range measurements, the frequency offset may be as high as $\pm 2.6 \times 10^{-10}$ for the UCT GPSDOs at $\tau_{PLL} = 150\,\text{s}$, and is dependent on the short-term GPSDO frequency accuracy at the time of recording. The average drift over 120 seconds is negligible at a few parts in 10^{-14} for all cases which seems to support the assumption that the average frequency offset is constant over short integration periods of a few seconds or less. As expected, and due to the lack of correlated cancellation, the bistatic carrier frequency is significantly noisier than the monostatic. Moreover, in comparison to the monostatic, the bistatic frequency noise and frequency spurs are higher by roughly a factor of 24 and ten, respectively. However, LOS phase compensation restores the bistatic fractional frequency to monostatic levels. It is also interesting that the frequency spikes present in the monostatic are absent in the phase compensated bistatic signal.

In summary, the frequency offset, noise and spurs originating from the $2.4\,\text{GHz}$ synthesiser sets a limit to the performance. The average bistatic frequency drift over the 120-second measurement is negligible. LOS phase compensation appears highly effective and reduces the bistatic frequency offset and noise to monostatic levels while cancelling the spurs entirely. However, more measurements of longer duration are required.

6.3.7   Doppler Phase Noise

This section compares the pulse-to-pulse Doppler phase noise for the monostatic, bistatic, and LOS phase compensated bistatic cases. The Doppler phase noise was estimated by calculating the PSD of the pulse-to-pulse matched filtered phase of the Roman Rock returns and scaling it to units of dBc/Hz. A Hanning window was applied before computing the PSD. The phase noise was computed using Matlab, but the algorithm was verified against the Stable32 software suit [99]. Note that the Doppler phase noise is very different from the IF phase noise and cannot be compared to the predictions in Section 3.6.5. Firstly, it is matched filtered. Secondly, the radar PRF of 1 kHz sets the maximum frequency offset of the Doppler phase noise to 500 Hz. However, the noise spans the entire single-sideband radar bandwidth of 25 MHz causing the upper Nyquist regions to fold back into the first. An IF phase noise measurement, on the other hand, is usually hardware band limited to prevent such aliasing. Moreover, there was never an opportunity to measure the NetRAD IF phase noise for the monostatic and quasi-monostatic cases, and it is left to future work. However, the Doppler phase noise is still thought to be insightful when comparing the monostatic, bistatic, and LOS phase compensated bistatic cases.

Figure 6.11 compares the Doppler phase noise of the monostatic, bistatic and LOS phase compensated bistatic returns of Roman Rock. The noise is aliased, and the slopes are not representative of the underlying noise processes. However, it is the perceived Doppler phase noise at a PRF of 1 kHz. The monostatic node has predominantly White PM noise above 40 Hz, and White FM noise below 40 Hz. For the bistatic case, the White FM noise starts to dominate at below 10 Hz. Above a 100 Hz, the bistatic phase noise is approximately 20 dBc/Hz higher than the monostatic. However, below 10 Hz the bistatic is only 10 dBc/Hz higher than the monostatic down to about 1 Hz. Bistatic LOS phase compensation only improves the phase noise for frequencies above 10 Hz by seemingly lowering the White PM level. However, LOS phase compensation lowers the bistatic phase noise for frequencies above 100 Hz to the monostatic level. However, White FM noise starts to dominate below 100 Hz. Below about 30 Hz the compensated bistatic is 10 dBc/Hz higher than the monostatic down to about 1 Hz. At below 8 Hz, the bistatic and LOS phase

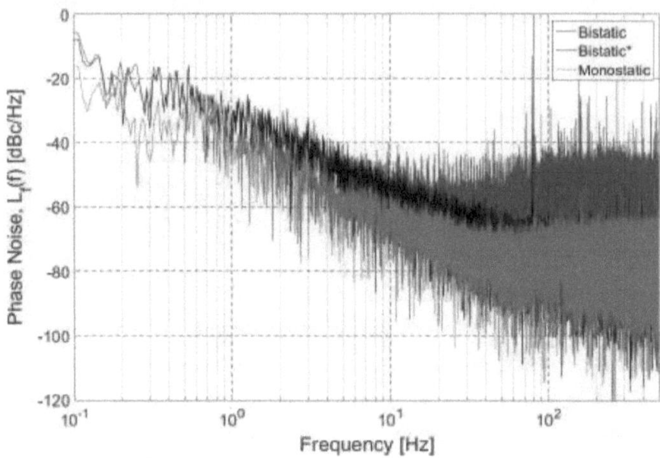

Figure 6.11: Close-In IF Phase Noise. The difference between the monostatic and bistatic is not nearly as large as predicted, and both exhibit White FM at the lower frequencies. This White FM noise suggests that there is a 10 dBc/Hz per decade noise cancellation in both cases. LOS phase compensation lowered the White PM noise to the monostatic level for frequencies above 40 Hz.

compensated bistatic are at a similar level.

In summary, LOS phase compensation lowered the White PM level by 20 dBc/Hz for frequencies above 40 Hz restoring the bistatic phase noise to the monostatic level. However, the bistatic phase noise for frequencies below 15 Hz is unaffected by LOS phase compensation.

6.3.8 Doppler Performance

This section compares the Doppler performance of the monostatic, bistatic, and LOS phase compensated bistatic signals. First, the Roman Rock Doppler profiles are examined. Then, the range-Doppler plots of the bistatic and LOS phase compensated bistatic Simon's Bay geometry are compared. Additionally, range-Doppler plots of the bistatic and the LOS phase compensated bistatic returns of a circling speedboat are compared.

An FFT spanning the entire 130-second recordings were used to compute the Doppler profiles and range-Doppler plots in this section. This unrealistically long

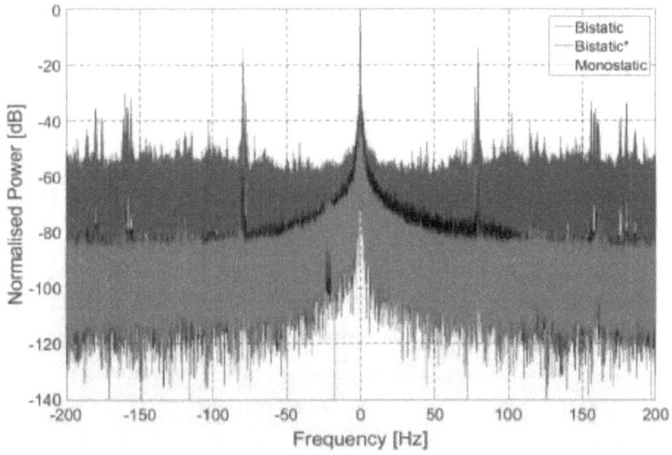

Figure 6.12: Roman Rock Doppler Profile. The Doppler profiles of the monostatic, bistatic and LOS phase compensated bistatic returns are compared. The bistatic noise floor is about 30 dB higher than the monostatic. However, LOS phase compensation restored the SCV to monostatic levels.

Doppler integration results in a very high Doppler resolution and also exaggerates the GPSDO phase dynamics. Hence, it is convenient to demonstrate the effect of LOS phase compensation on the bistatic Doppler performance.

The Roman Rock Doppler profiles for the monostatic, bistatic and LOS phase compensated bistatic returns are compared in Figure 6.12. At Doppler frequencies above ± 50 Hz, the bistatic noise floor is approximately 30 dB higher than the monostatic resulting in a much lower bistatic SCV. The bistatic spurious Doppler frequencies are also more pronounced. Moreover, the Doppler frequency spurs at ± 80 Hz are roughly 12 dB higher for the bistatic case. LOS phase compensation lowers the bistatic noise floor to monostatic levels. Further, the compensated Doppler frequency spurs are broadened but suppressed to 4 dB below that of the monostatic. Interestingly, the compensated bistatic is weighted unsymmetrically to the positive frequencies, while the monostatic appears symmetrical. This is believed to be caused by the waves travelling into the bay in the radial direction of the bistatic node while being perpendicular to the monostatic node. However, an important point here is that LOS phase compensation sufficiently restored the bistatic SCV to uncover this wave motion in the presence of the large Roman Rock return. This wave motion would have been otherwise undetectable.

Figure 6.13 compares the range-Doppler plots for the bistatic and LOS phase compensated bistatic returns for the Simon's Bay geometry. Roman Rock and both the transmitter and receiver platforms are static. Ideally, both the breakthrough and Roman Rock should show minimal Doppler spreading. However, the bistatic case in Figure 6.13a show significant Doppler spreading at both the breakthrough and Roman Rock due to the higher bistatic phase noise. This reduced SCV would make small targets difficult to detect within the Roman Rock range bin. Further, the pronounced Doppler spurs appear as 'tramlines' at frequency multiples of $\pm80\,\text{Hz}$. The Doppler spreading is significantly reduced for the LOS phase compensated case in Figure 6.13b. The overall lower noise level improves the dynamic range. However, the first harmonic Doppler spurs at $\pm80\,\text{Hz}$ are still visible.

Figure 6.14 shows the bistatic RTI plot and associated range-Doppler plots of a circling speedboat. This recording had a different geometry with a baseline of a few hundred metres. In the RTI plot in Figure 6.14a, the breakthrough is visible at range bin 196, and the circling speedboat is visible at a bistatic range of $3714\,\text{m}$. The speedboat completes roughly 2.5 revolutions at approximately 45 seconds per revolution. There are also a few static targets in the foreground. The breakthrough and other static targets show some range ripple because the GPSDO PLL locking transients had not yet decayed entirely at the time of the recording. However, this slow phase variation appears to have a more significant effect on the range than on the Doppler response of interest. The bistatic range-Doppler plots for the bistatic and LOS phase compensated bistatic cases are compared in Figures 6.14b and 6.14c, respectively. The uncompensated range-Doppler plot shows a relatively good Doppler response for the speedboat with little spreading, albeit in the absence of larger static targets at the same range. However, the breakthrough has the familiar Doppler spreading and 'tramlines'. The smaller static foreground targets have a less pronounced Doppler spreading. The hazy positive Doppler response appearing from range bin 650 and onwards is presumably wave clutter Doppler, offset from zero by the wave breaking direction. As before, the LOS phase compensation significantly reduced the breakthrough Doppler spreading as well as the overall noise floor.

In summary, the bistatic SCV and Doppler spurs are roughly $20\,\text{dB}$ worse and $12\,\text{dB}$ higher than the monostatic, respectively. Large bistatic targets show signific-

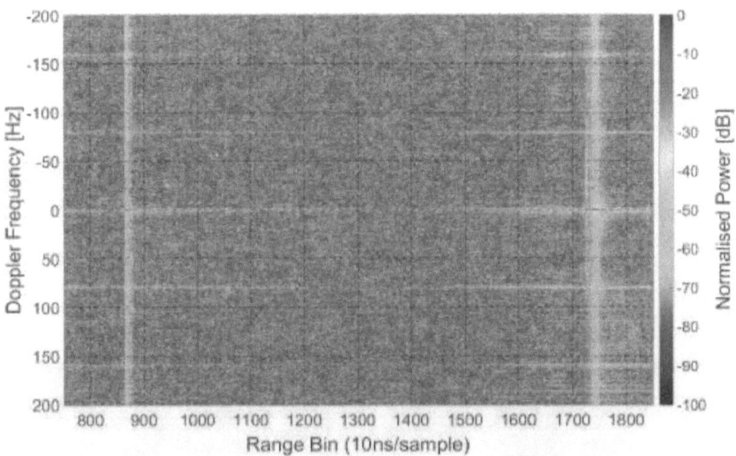

(a) Bistatic Range-Doppler Plot: Node32

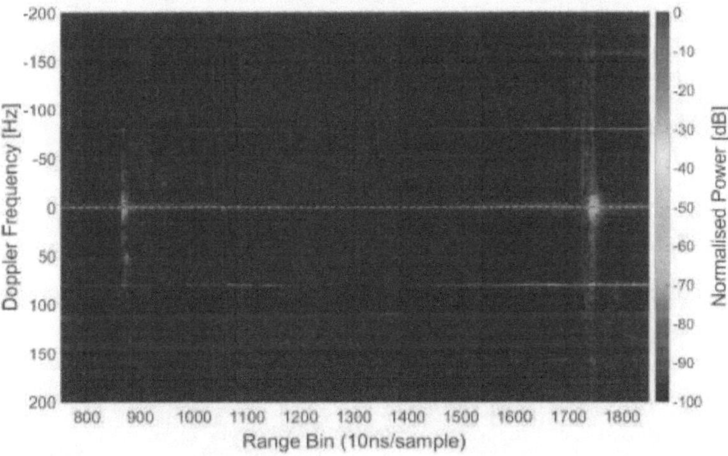

(b) Bistatic Range-Doppler Plot: Node32 (after line-of-sight phase correction)

Figure 6.13: Bistatic Range-Doppler Plot of the Simon's Bay Geometry. These plots show the breakthrough and Roman Rock returns before and after line-of-sight phase correction. The phase of the direct breakthrough was used as the phase correction reference.

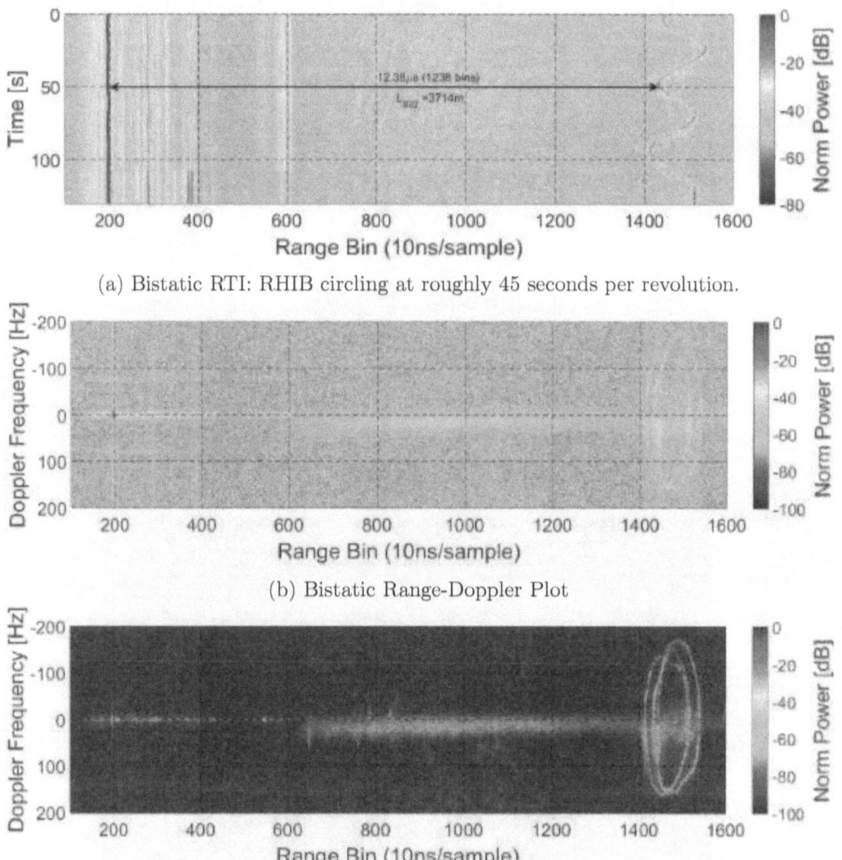

(a) Bistatic RTI: RHIB circling at roughly 45 seconds per revolution.

(b) Bistatic Range-Doppler Plot

(c) Bistatic Range-Doppler Plot (after line-of-sight phase correction)

Figure 6.14: Bistatic Range-Time Intensity and Range-Doppler Plots of a Circling Speed Boat. The boat was circling at roughly 45 seconds per revolution at a bistatic range of 3.7 km. The phase of the direct breakthrough was used as the phase correction reference. The hazy positive Doppler response appearing from range bin 650 and onwards is presumably wave clutter Doppler, offset from zero by the wave breaking direction.

ant Doppler spreading and spurs appearing as 'tramlines'. However, in the absence of large targets, the bistatic Doppler performance of small targets appears reasonable. LOS phase compensation lowers the SCV and Doppler spurs to monostatic levels. Thus, breakthrough compensation used in combination with low-cost quartz GPSDOs makes affordable high-performance short baseline bistatic Doppler systems possible at minimal additional processing. However, it requires that both the breakthrough and target be recording within a single pulse.

6.3.9 Non-Coherent Integration (linear range creep)

This section investigates non-coherent pulse integration in the presence of a linear range advance caused by a bistatic frequency offset. This effect was discussed in Section 3.4.1. Moreover, the theory predicts that the normalised amplitude gain would decrease, the target 3 dB main lobe would widen, and that the target range would advance for increasing range creep.

The data in Figure 6.6 is used to test this theory. In this figure, the frequency is offset by 4.269×10^{-7} at 57 seconds to induce a false Tx-Rx velocity. After heavy zero-padding, it takes the peak amplitude response of the direct breakthrough roughly 11 pulses to advance by its 3 dB main lobe width. However, there is some signal noise, and the result is only approximate for multiples of 11 pulses. Also, sidelobes of nearby scatterers make it difficult to find a good dataset.

Figure 6.15 plots the normalised non-coherent amplitude gain for various values of normalised range creep. Moreover, the frequency offset data in Figure 6.6 is non-coherently integrated for integer multiples of 11 pulses or b. For the case of $b = 0$, 11 pulses from the static section in Figure 6.6 before 57 seconds is integrated. As expected, the normalised amplitude decrease, the 3 dB main lobe broadens, and the range advances for increasing integration periods. However, the range advance in the real data is not perfectly linear, and trends do not exactly follow the predictions in Figures 3.10 to 3.12. Further investigation with more data is required to determine how well the real world compares to the theory.

6.3.10 Coherent Integration (linear phase creep)

This section investigates coherent pulse integration in the presence of a linear phase advance caused by a bistatic frequency offset. This effect was discussed in Sec-

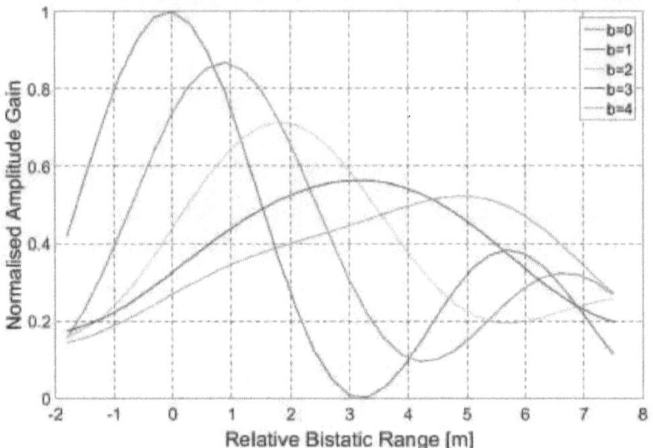

Figure 6.15: Normalised Non-Coherent Amplitude Integration Gain Versus the Total Range Creep. The normalised amplitude gain decreases, the main lobe widens, and the target range advances for increasing range creep. Clutter adjacent to the main lobe causes the oscillating behaviour and makes this effect difficult to prove using real data.

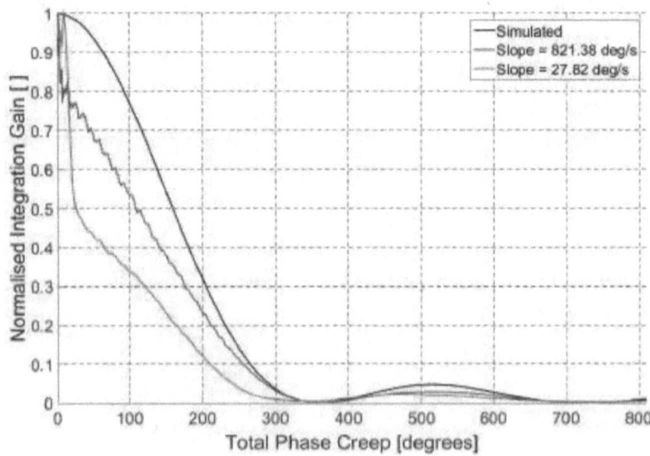

Figure 6.16: Normalised Coherent Integration Gain Versus Total Phase Creep. The integration requires about 16 pulses to converge and causes an initial spike in the normalised gain. The data with the slope of 821.38 °/s contain less 1/f noise and follows the simulated plot more closely. The 27.82 °/s data is integrated for longer and contains more 1/f noise.

tion 3.4.2. Moreover, the theory predicts a diminishing coherent SNR gain for increasing values of total phase creep.

Two datasets where the breakthrough had linear phase slopes were used to test this theory. The phase slopes of the data sets were $821.38\,°/s$ and $27.82\,°/s$ equating to constant frequency offsets of 9.5×10^{-10} and 3.22×10^{-11}, respectively. For each dataset, a zero-drift control was created such that ideal and non-ideal coherent integration can be compared using similar data. Moreover, a linear phase fit is applied to each dataset whereafter the LOS phase compensation algorithm of Section 6.3.5 is used to remove the phase slope in each case. Both the zero-drift and linear drift datasets are then coherently integrated and the SNR loss caused by the phase drift calculated using (3.57).

The normalised coherent integration gain and coherent SNR loss are plotted in Figures 6.16 and 6.17, respectively. The two frequency offset datasets are compared to the simulated case of pure white integrated noise. Note that a total phase creep of $810°$ equates to one second and 29 seconds of integration for the phase slopes $821.38\,°/s$ and $27.82\,°/s$, respectively. From Figure 6.11, the coherent integration spans into the $1/f$ noise region for both cases. However, the $27.82\,°/s$ dataset is integrated significantly longer and is expected to contain more $1/f$ noise. In both cases, the normalised integration gain loosely follow the ideally simulated trend. However, the integration requires about 16 pulses to converge and causes an initial spike in the normalised gain. As expected, the $821.38\,°/s$ dataset converges more quickly than the $27.82\,°/s$ dataset. Further, it is thought that the $821.38\,°/s$ dataset follows the simulated plot better because it contains a larger component of white noise. Figure 6.13b plots the coherent SNR loss for a total phase creep of between $0°$ and $266°$. The $821.38\,°/s$ dataset follows the simulated trends closely. However, the $27.82\,°/s$ dataset with more $1/f$ noise deviates somewhat and reaches an SNR loss of $10\,dB$ after about $250°$. The $1/f$ noise component presumably attenuates the SNR loss peaks at multiples of $360°$. Moreover, the peaks simulated for white noise strive to infinity. However, the peak loss at $360°$ is $38\,dB$ and $20\,dB$ for the $821.38\,°/s$ and $27.82\,°/s$ datasets, respectively.

In summary, the coherent integration of the real data compares well with the theory developed in Section 3.4.2. This theory assumes pure white integrated noise. However, both datasets contain a $1/f$ noise component. The coherent SNR loss of the $821.38\,°/s$ which is integrated for a shorter period follows the simulated plot closely. However, the $27.82\,°/s$ dataset with more $1/f$ noise deviates somewhat. Moreover,

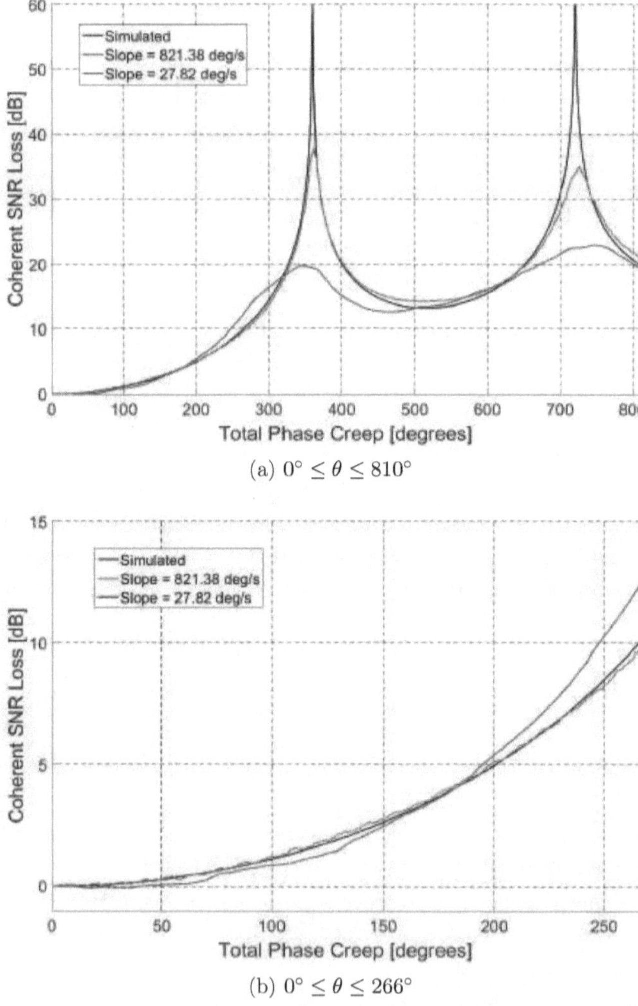

(a) $0° \leq \theta \leq 810°$

(b) $0° \leq \theta \leq 266°$

Figure 6.17: Coherent SNR Loss Versus Total Phase Creep. Up to approximately 200°, both datasets follow the simulated plot closely. The $1/f$ noise components appear advantageous since it attenuates the SNR loss peaks located at multiples of 360°.

a larger $1/f$ noise component is possibly advantageous since it attenuates the SNR loss peaks located at multiples of 360°. However, refer to Carotenuto et al.[106, 107] for more on the effects of phase noise on radar signal processing which is out of the scope of this relatively rudimentary analysis.

6.4 Conclusion

This chapter analysed the performance of the one-way GPSDO synchronised Net-RAD. First, it described how NetRAD was converted from a cabled system to an entirely wireless network radar. A tristatic experiment was set up in Simon's Bay, South Africa. This geometry had a maximum baseline of 2.3 km, and the maximum bistatic range to Roman Rock was 3.4 km. The transmitter had an unobstructed LOS to the two passive bistatic nodes. Then, the bistatic time, phase, frequency, Doppler phase noise and the Doppler performance of the GPS synchronised Net-RAD was measured. The monostatic, bistatic, and LOS compensated bistatic cases were compared. Finally, the effect of a constant bistatic frequency offset on both non-coherent and coherent integration is studied. Moreover, the real data was compared to the models developed in Section 3.6. The UCT GPSDOs were tuned to a suboptimal $\tau_{PLL} = 150$ s since longer time constants proved impractical. The quick-locking filter described in Section 4.3.4 was not yet fully developed at the time.

Time precision could not be measured without independently surveyed site positions and requires further investigation. The time accuracy was measured using the change in the breakthrough time of arrival across a static baseline. The maximum measured bistatic time drift of ±5 ns about the mean over a 20-minute period is consistent with the ±5.5 ns zero-baseline calibration for $\tau_{PLL} = 150$ s. Operating at the optimal $\tau_{PLL} = 1000$ s should improve the time accuracy to ±3 ns. Hence, a short baseline sub-10 ns time synchronisation may be possible under ideal circumstances but will require the equipment offsets to be calibrated, professionally surveyed co-ordinates [31], and temperature stabilised GPS antennas [61]. However, in practice, one would expect larger time biases and one-way quartz GPSDO time transfer to achieve a sub-20 ns performance. Common-view or carrier-phase GPS time transfer will generally be required for sub-10 ns timing. However, for precisely known baselines, of a few kilometres, it is probably cheaper and more practical to use the breakthrough to achieve sub-10 ns time synchronisation. However, measurements with known site positions are required to confirm.

In these limited measurements, the low-cost quartz GPSDOs, tuned to a sub-optimal $\tau_{PLL} = 150\,$s, proved adequate to achieve both NetRAD requirements of a linear phase during pulse integration and a frequency offset below 10^{-10}. The GPSDO phase steering caused the bistatic phase to drift during the 130-second recordings. However, the slopes were linear over a few seconds or less. The maximum measured phase slope was -92$\,^\circ$/s to achieve a sub-1 dB coherent SNR loss. This slope translates to a frequency offset of 1.1×10^{-10}, or a Doppler error of 0.256 Hz at 2.4 GHz, or a radial target velocity error of 0.115 km/h. The averaged frequency drift over a 120-second recording was a few parts in 10^{-14} supporting the assumption of a constant bistatic frequency offset over a few seconds or less. However, note that the relative GPSDO frequency slowly drifts over minutes and the measured Doppler phase slope may fall anywhere within the measured zero-range short-term frequency range of $\pm 2.6 \times 10^{-10}$ at $\tau_{PLL} = 150\,$s. Nonetheless, one expects the UCT GPSDOs to easily achieve the phase and frequency requirements when their low-noise OCXO buffers are fitted. These buffers are expected to lower the frequency inaccuracy by a factor of 10. However, the 2.4 GHz PLL-synthesisers limit both the mono- and bistatic frequency offsets. This highlights that the subsystem performance should match that of the STALO.

The Doppler phase noise at a PRF of 1 kHz was computed, and the monostatic, bistatic, and LOS phase compensated cases were compared. The monostatic node has predominantly White PM noise above 40 Hz, and White FM noise below 40 Hz. For the bistatic case, the White FM noise starts to dominate at below 10 Hz. Above a 100 Hz, the bistatic phase noise is approximately 20 dBc/Hz higher than the monostatic. LOS phase compensation partially cancelled the bistatic Doppler phase noise above 10 Hz reducing it to monostatic levels at above 100 Hz. There was never an opportunity to measure the mono- and quasi-monostatic IF phase noise NetRAD of NetRAD for comparison to the predictions in Section 3.6.5. This is left to future work.

The bistatic SCV and Doppler spurs are roughly 20 dB worse and 12 dB higher than the monostatic, respectively. Large bistatic targets show significant Doppler spreading and spurs appearing as 'tramlines'. However, in the absence of large targets, the bistatic Doppler performance of small targets appears reasonable. LOS phase compensation improved the range-Doppler plots to near monostatic levels.

However, the first harmonic 'tramlines' were still visible.

Finally, the effect of a constant bistatic frequency offset on both non-coherent and coherent integration was studied using real bistatic data. In both cases, the real data compared well with the theory of Section 3.4. A larger 1/f integrated noise component appeared to be advantageous under certain conditions for the coherent case. However, more data and further analysis are required.

In summary, low-cost quartz GPSDOs can synchronise NetRAD to within its phase and frequency requirements comfortably. However, range accuracy will be limited to less than twice its 3 m range resolution. As expected, the bistatic phase returns are significantly noisier than that of the monostatic reducing the SCV by 30 dB. However, LOS phase compensation used in combination with low-cost quartz GPSDOs proved to be an affordable solution to high-performance short baseline bistatic systems. Moreover, the phase across a static baseline was used to restore the bistatic phase, frequency, phase noise and Doppler performance to near monostatic levels. It was further confirmed that near identical results are achieved when Roman Rock, a large common static target, is used as the phase reference. The results in this chapter are based on limited data and short 120-second recordings. Thus, more data and further analysis are required using GPSDOs that are optimally tuned and buffered.

Chapter 7

Conclusion, Summary and Discussion

This text evaluated the feasibility of synchronising a coherent pulsed-Doppler networked radar, using low-cost quartz GPSDOs across baselines of a few kilometres. It further quantified the improvement in bistatic phase coherence when using LOS phase compensation in combination with GPSDO synchronisation.

An imperfectly synchronised bistatic NetRAD Tx-Rx pair was modelled directly relating the bistatic range, Doppler and phase errors to synchronisation performance. This model was used to define the NetRAD synchronisation requirement based on the desired radar performance. Then, the UCT GPSDO was designed with the goal of being open, versatile, extensible, and with a known and predictable behaviour to study network radar synchronisation. These GPSDOs are capable of both rapid phase-lock and precise network-wide time synchronisation. Their performance was carefully calibrated at a zero baseline to enable meaningful comparisons to real bistatic measurements across larger baselines. A tristatic experiment was set up in Simon's Bay, South Africa and the UCT GPSDOs were used to synchronise the tri-node NetRAD across baselines of up to 2.3 km. Strong bistatic reflections from the Roman Rock lighthouse were recorded. LOS phase compensation was applied to this data and compared to that of the monostatic and bistatic returns. Moreover, the phase, frequency, and Doppler performance of the monostatic, bistatic, and LOS compensated bistatic Roman Rock returns are compared.

In the remaining sections, this **book** is concluded by applying the main findings to the principal hypothesis and associated research questions. Then, follows a broader chapter by chapter summary listing the novel contributions while discussing the validity and limitations of the research design, and future research.

7.1 Conclusion

Low-cost quartz GPSDOs, based on one-way GPS time transfer, are suitable to synchronise coherent, pulsed-Doppler, networked, radars with carrier frequencies of a few gigahertz and moderate bandwidths of tens of megahertz across short baselines of a few kilometres. LOS phase compensation can enhance the bistatic phase coherence of such a GPSDO synchronised bistatic radar to near monostatic performance levels.

The bistatic time accuracy determines the possible range accuracy. At zero-range and identical coordinates, the initial time bias of commercial low-cost GPS receivers can be pre-calibrated to within 1 ns corresponding to a bistatic range error of 0.3 m. However, this error could be tens of nanoseconds larger when using GPS auto-surveyed coordinates [31]. Also, ambient temperature differences between the sites may contribute another 2 ns to 3 ns (or 0.6 m to 0.9 m) of uncertainty [61]. However, these values were never corroborated during field measurements which would have required precisely and independently surveyed antenna coordinates. It is nonetheless conceivable from [60] that sub-10 ns (sub-3 m) uncertainties could be achieved using precisely surveyed coordinates and temperature stabilised equipment. The UCT GPSDO zero-range time drifted, consistent with [60], by less than ± 5.5 ns (± 1.65 m) and ± 3 ns (± 0.9 m) for $\tau_{PLL} = 150$ s and $\tau_{PLL} = 1000$ s, respectively. This was verified across baselines of a few kilometres for $\tau_{PLL} = 150$ s. Hence, an initial range offset of below 3 m plus a range drift of ± 0.9 m can be expected. If the requirement is to synchronise to within the radar range resolution, then low-cost GPS is good for radar bandwidths of below 37.5 MHz.

Most GPSDOs easily reach long-term averaged frequency offsets and drifts of a few parts in 10^{-13} and 10^{-14}, respectively. However, it is the instantaneous frequency offset at the time of Doppler measurement that determines the possible radar target velocity accuracy. Moreover, the instantaneous GPSDO frequency accuracy is typically orders of magnitude worse than the long-term average and may range between $\pm 7 \times 10^{-10}$ and $\pm 2 \times 10^{-9}$ [81]. Nonetheless, a radial target velocity accuracy of below 1 km/h requires $y_o \leq 9.27 \times 10^{-10}$ which is readily achieved by many low-cost GPSDOs. At zero-baselines, the UCT GPSDO achieved $\pm 2.6 \times 10^{-10}$ and $\pm 1.3 \times 10^{-10}$ at $\tau_{PLL} = 150$ s and $\tau_{PLL} = 1000$ s, respectively. The field measured bistatic frequency offsets were all within $\pm 2.6 \times 10^{-10}$ but more and longer-term measurements are desirable.

The frequency drift during a Doppler measurement must be less than the fractional Doppler resolution and is determined by the carrier frequency and integration time. This drift requirement gets more stringent for higher carrier frequencies and longer integration times. For a one-second integration period and maximum integration loss of 1 dB, frequency drifts of less than 2.1×10^{-10}, 5×10^{-11}, and 2.5×10^{-11}, are required for carrier frequencies of 2.5 GHz, 10 GHz, and 20 GHz, respectively. Phase-locked GPSDOs drift negligibly during short periods of a few seconds or less. Hence, the above frequency drift requirements are readily achievable by low-cost GPSDOs.

The allowable phase drift during coherent integration may be expressed as a maximum allowable time drift as a function of the carrier frequency. However, a constant frequency offset is often the dominating contributor to time drift simplifying this requirement to an allowable frequency offset. The coherent integration gain loss due to a linear phase drift was derived and later experimentally verified. Moreover, a total linear phase drift of 95°, 160°, and 266° results in a coherent integration gain loss of 1 dB, 3 dB, and 10 dB, respectively. For a one-second integration period, frequency offsets of below 1.1×10^{-10}, 2.6×10^{-11}, and 1.3×10^{-11} are required at carrier frequencies of 2.4 GHz, 10 GHz, and 20 GHz, respectively. The UCT GPSDOs with a measured zero-baseline $y_o \leq \pm 1.3 \times 10^{-10}$ at $\tau_{PLL} = 1000$ s should maintain a coherent integration losses of 1 dB, 3 dB, and 10 dB, for carrier frequencies of 2 GHz, 3.4 GHz, and 5.6 GHz, respectively. During bistatic field measurements at $\tau_{PLL} = 150$ s, a sub-1 dB integration loss was maintained. However, this is expected to deteriorate when doing more and longer recordings.

The lack of IF phase noise cancellation at the bistatic receiver has a detrimental effect on the SCV. The NetRAD bistatic phase noise was never measured, but from [10] it is estimated to be approximately 80 dBc/Hz worse than that of the monostatic at 1 Hz for a target at 3 km. Moreover, during bistatic field measurements, the bistatic SCV was measured to be 30 dB worse than that of the monostatic. Further, the bistatic spurious frequencies were about 12 dB more pronounced than that of the monostatic. This lack of IF phase noise cancellation is one of the biggest challenges when synchronising bistatic radar using standalone GPSDOs.

Lowering the STALO close-in phase noise and spurious to improve the SCV is costly and often unattainable. Hence, the use of LOS phase compensation proved promising. Moreover, during the bistatic field measurements, LOS phase compensation improved the bistatic phase, frequency, and Doppler performance to near monostatic levels.

Thus, low-cost GPSDOs are in many cases sufficient to synchronise bistatic radar. Nevertheless, the bistatic performance remains limited by the inevitable Tx-Rx phase dynamics, frequency inaccuracy, and relatively poor SCV. The above mentioned may be improved by using more expensive GPSDOs containing more stable and lower phase noise STALOs. However, LOS phase compensation has the potential to improve the performance of very cheap GPSDOs to that of near monostatic levels over baselines of a few kilometres. GPSDO pre-synchronisation is still required to synchronise the Tx-Rx PRF and triggering, but poor phase dynamics and phase noise can be corrected during real-time postprocessing using the direct breakthrough as reference.

7.2   Summary and Discussion

This section contains a chapter by chapter summary. It simultaneously discusses the research implications and limitations and recommends future research. The contributions contained in each chapter are listed using bullet points.

Chapter 1, started with a brief background review and later defined the primary hypothesis with the associated research questions. Then, it listed the main contributions produced by this work, followed by a concise book outline.

Chapter 2, reviewed the literature related to the time, frequency, and phase synchronisation of bistatic, multistatic, and networked radar systems. It examined how the bistatic time, frequency, and phase synchronisation relate to the specification of bistatic STALOs. Few publications discussing the time, frequency, and phase specification o f b istatic r adar f rom a s ystems p erspective w ere f ound. A lso, such publications often assumed that the time and frequency offsets, and oscillator ageing are zero where the time and phase synchronisation are functions of frequency stability, $\sigma_y(\tau)$, only. This assumption is overly simplistic, and the relative frequency offset may often dominate. Further, it remains difficult for the bistatic design engineer to quickly select a synchronisation technology based on the desired pulsed-Doppler performance. Hereafter, alternative synchronisation methods were compared which included synchronisation via RF and optic fibre c ables, a cross RF or optical free-space links, and using the direct or reflected r adar e missions. The various GPS time transfer techniques, their implementation and performances were reviewed at length. GNSS time transfer was identified a s p otentially s uitable to ground-based bistatic radar. It is cheap, simple, autonomous, passive and covert, and does not require direct LOS. The focus remained on low-cost one-way GPS time transfer. Then, followed a state-of-the-art analysis of GPSDO technology. This included the published literature on GPSDO architecture, feedback control strategies, time synchronisation techniques, and achieved performance. There exists a substantial opportunity for innovation within GPSDO feedback control strategies. Areas requiring research include algorithms capable of auto-tuning the feedback bandwidth based on the reference performance, rapid-locking feedback loops, and hold-over algorithms. There is also a need for precise time synchronisation circuits that are in phase with the STALO. A rapid development environment for feedback and filtering algorithms that could be tested with hardware in the loop is highly desirable. Commercial GPSDOs may behave very differently, and manufacturer specifications are often only estimates of their true performances [31]. Manufacturer specifications are also not well suited to bistatic radar applications. Thus, GPSDOs used in bistatic applications, require extensive pre-calibration. Finally, the few published examples of GPS synchronised bistatic (or multistatic) were considered. The only recent examples of bistatic GPS synchronisation found was for VHF over the horizon radar. Further, the combined use of the sidelobe breakthrough phase to remove the inevitable GPSDO phase dynamics [27] appeared to be a promising solution to low-cost but high-performance synchronisation. The performance of this technique requires further investigation and quantification.

Chapter 3 expanded on the existing literature [3, 7–10, 20, 32] to better specify the synchronisation requirements of bistatic pulsed-Doppler radar. It modelled an imperfectly synchronised bistatic Tx-Rx pair with the purpose to predict the effect of imperfect synchronisation on an otherwise ideal bistatic system. The total accumulated relative phase error was lumped together at the receiver and expressed using a well-known, but simple, oscillator model. However, an arbitrary complex model could replace this model. The IF signal was expressed as a function of transmitter-target-receiver time delay, bistatic doppler frequency, and imperfect Tx-Rx synchronisation.

- This IF signal was sampled and expressed as a function of radar pulse number, ADC sample number, and an arbitrary future time when the first radar pulse is sampled.

- Then, the individual phase, Doppler, and radar range error contributions by the heterodyne conversion, radar PRF and ADC sampling were derived. The total error contribution is the sum of the individual error contributions.

- It was also shown that ADC sampling does not introduce a significant error over that of down-conversion for stable oscillators.

This model is sufficient to model the homodyne NetRAD operating at baseband. Future research may follow a similar approach to derive accurate models of other architectures, for example, superheterodyne and direct sampled systems. For directly sampled systems, the ADC sampling stage will contain the most significant error due to imperfect synchronisation.

- A set of plots directly relating oscillator parameters to radar performance was presented. These plots can be used to quickly gauge the Tx-Rx synchronisation demands based on the required bistatic performance.

However, these plots are based on the somewhat idealistic synchronisation model, and future research may include models which better describe specific circumstances and synchronisation techniques.

Bistatic pulse-Doppler radars are affected by the instantaneous Tx-Rx frequency offset at the time of measurement, and the frequency and phase drift during the measurement. In many cases, such as when using GPSDO synchronisation, the initial relative frequency offset dominates. Expressions were derived describing the effect of a constant Tx-Rx frequency offset on both non-coherent and coherent integration. The goal was not to make an in-depth study of bistatic pulse integration. The purpose was to enable the designer to specify the required synchronisation based on the desired integration performance.

- A large constant Tx-Rx frequency offset during non-coherent processing causes a range advance resulting in a reduction in amplitude gain, a broadening of the target main lobe, and a range error.

It is desirable to also quantify the loss in non-coherent integration gain due to a constant frequency offset. However, the non-linear nature of envelope detection makes this a rather difficult problem to solve.

- A constant Tx-Rx frequency offset during coherent processing causes a phase advance resulting in a loss of coherent integration gain. Moreover, it was shown that a total phase creep of 95°, 160° and 266° result in an SNR loss of 1 dB, 3 dB and 10 dB, respectively.

- A plot allowing the designer to select an appropriate synchronisation technology based on the carrier frequency and allowed coherent SNR loss (or phase creep) was presented.

Bistatic pulse integration in the presence of a constant Tx-Rx frequency offset is believed to be equivalent to the monostatic integration of a target travelling at a constant radial velocity. Hence, future research may investigate how range migration mitigation techniques could be applied to imperfectly synchronised bistatic processing.

It was demonstrated that the plots and derivations mentioned above could be used to specify bistatic synchronisation based on the required performance. The NetRAD radar was used as an example.

- Moreover, the desired NetRAD range, doppler, and integration requirements were specified as functions of time, frequency, and phase synchronisation. The range requirement was specified in terms of initial time offset and the time drift during the hold-over period. The frequency synchronisation requirement was based upon the required radial target velocity accuracy and the Doppler resolution. The phase specification was simplified by assuming that the constant Tx-Rx frequency offset dominates. Subsequently, the required maximum frequency offset was derived as a function of the tolerable SNR loss during integration and the integration time.

Using the derived derivations and plots, it is now possible to rapidly quantify and select an appropriate synchronisation technology based on the required system-level NetRAD performance. It was estimated that low-cost quartz GPSDOs are adequate to synchronise NetRAD. However, the desired range accuracy must be relaxed from one-tenth of the range resolution to just within the range resolution or a range bin of 3 metres. Also, it appears impossible for low-cost GPSDOs to provide the required ultra-low close-in phase noise to ensure an SCV equivalent to that of monostatic radar. Hence, a LOS phase compensation technique was proposed where the breakthrough along a static baseline is used to remove GPSDO induced phase dynamics.

Chapter 4 detailed the hardware, firmware and software design of the three low-cost quartz UCT GPSDOs.

- These unique GPSDOs were designed to serve as a platform for bistatic synchronisation research. They are open, versatile, extensible, with predictable behaviour. Also, the user has real-time read and write access to the low-level subsystems. This text focussed on finding solutions to precise network-wide time synchronisation, reducing the GPSDO lock-in time, and developing a hardware in the loop GPSDO development environment [34].

The UCT GPSDO uses a built-in low-cost Motorola M12+ GPS timing receiver and a moderately priced Oscilloquartz OCXO 8788 STALO. The phase detector has a resolution of 65 ps and jitter of 50 ps with a measurement range of 1 µs. The STALO steering is done using a 20-bit DAC. The clock distribution circuitry includes integer digital dividers and delivers the outputs in both single-ended and differential logic. An external PPS input makes it possible to synchronise to alternative lower jitter PPS sources. These GPSDOs performed well and were adequate to collect meaningful multistatic sea clutter and small sea vessel data [27, 100–102, 104, 105, 108]. However, due to a lack of time and funding the GPSDO hardware never made it past the first generation prototype stage. Consequently, the frequency steering circuitry is over complicated and could be much improved. Also, due to the lack of time, the discrete common-base BJT output buffer amplifiers were never populated. Instead, the MMIC RF amplifiers, meant for testing only, are still in use. These amplifiers offer much less isolation and increase the phase noise by 15 dBc/Hz to 20 dBc/Hz. Hence, it is strongly recommended to populate the discrete BJT amplifiers for future measurements. Further, a redesign of the steering and clock distribution circuitry should include lower noise amplifiers, better isolation and buffering, and better filtering.

- The UCT GPSDO firmware is entirely FPGA-based and has a parallel multi-microprocessor architecture using the Xilinx picoBlaze microprocessor. It is an IRQ based system where the microprocessors communicate exclusively via a common RAM space. The multiple subsystems are autonomous and run in parallel which alleviates the timing constraints present within a fully sequential system. The system is easily extensible whereas fully VHDL-based systems are difficult to modify. Additional functionality and microprocessors can be ad-hoc added without affecting the existing VHDL code. This parallel architecture was independently developed, and the application to GPSDO firmware is believed to be unique. (Section 4.2.1)

Various pBlaze blocks could be implemented to enhance the autonomous functioning of the UCT GPSDO. Some of this functionality may include outlier detection, phase-locked detection, an improved and quick-locking PLL, and better support for GPS communication via the GPSDO serial port.

- The UCT GPSDO has a novel time synchronisation mechanism, called the EP. This EP can establish a precise network-wide time epoch derived from the GPSDO STALO. This low jitter GPS-time synchronised rising edge can be produced at an arbitrary, and user-specified, future time. This mechanism selects the OCXO rising edge which best estimates the true GPS time. [35] Thus, the EP is in-phase with the OCXO output by design. This epoch pulse mechanism is believed to be unique. (Section 4.2.4)

The EP mechanism has been used to time synchronise the NetRAD nodes over baselines of up to 5 kilometres. It achieved a zero-baseline GPSDO-to-GPSDO time synchronisation error of sub-10 ns which corresponds to the expected single-carrier one-way GPS time transfer performance.

- A PC-based GUI was developed which can control and monitor the UCT GPSDO in real-time via the serial interface. This software includes a smart PLL filter object. When this software filter is enabled, the sawtooth corrected PLL phase errors are read from the GPSDO and filtered in software. The result of the filter output is then written directly to the GPSDO's frequency steering DAC. Thus, the PC-based PLL filter and associated algorithms are tested in real-time with the GPSDO hardware in the loop. This feature allows the rapid development and testing of sophisticated filters and algorithms. As a demonstration, a sophisticated filter algorithm with outlier removal, phase-locked detection, and a quick-locking PLL filter was developed and tested [35]. (Section 4.3.1)

Future uses may include the study of different hold-over techniques in GPS-denied environments and auto-tuning filters. Moreover, an LSF pre-filter could be added to smooth the input to the loop filter [64]. The PLL-based design may be changed to a more versatile Kalman-based design capable of improved hold-over and temperature compensation through oscillator modelling [64, 67]. It is also desirable that the GPSDOs adapt automatically to the quality of the PPS reference input. [68] Finally, the software framework could be extended to have multiple GPSDOs co-operate across a TCP/IP network. Then, common-view GPS time transfer may be implemented in real-time, or the GPSDOs may function as a distributed clock where the collective steering is processed centrally.

- A phase-locked detection algorithm, tailored to the high jitter GPS PPS input, was designed and demonstrated. The algorithm is insensitive to PLL filter time constant when the size of the moving window is kept equal to the time constant. Examples of GPSDO phase-locked detection algorithms are very sparse in the literature. This algorithm is believed to be unique. (Section 4.3.3)

The phase-locked detection parameters were matched empirically to the jitter of the sawtooth corrected GPS PPS reference but adapts to the time constant of the critically damped locking transient. Future work may involve an algorithm which also auto-adapts to the reference input noise and PLL damping factor.

- A novel quick-locking PLL was designed and demonstrated. This adaptive PLL reduced the lock-in time by a factor of four. It does this by ramping the PLL filter time constant in real-time while simultaneously keeping the PLL critically damped during phase acquisition. [35] This filter is considered to be unique. (Section 4.3.4)

Future work may include adding the capability to adjust the damping factor in real-time. This quick-locking scheme could also be compared to other architectures such as the F-PLL with continuous frequency assist [80], or the two-step approach described by Baojian, Dehai, and Dazhi [30].

Some of the time synchronisation mechanisms described in the literature rely on special variable delay line [30] or less common DDS frequency steering GPSDO architectures [22, 72]. Similarly, some quick-locking mechanisms rely on architectures where the frequency and phase could be tuned indepedendly. However, both the EP and quick-locking mechanisms demonstrated in this text could be implemented on existing DAC-steered GPSDO architectures without any (or minimal) hardware modifications.

Chapter 5 measured the zero-baseline relative synchronisation performance of the UCT GPSDOs. The purpose was to produce three well-calibrated clocks, verify their correct and predictable behaviour, and to verify that their PLL bandwidths are optimally tuned. Their relative time (phase), frequency, frequency stability, and phase noise performances were measured at zero-baselines. Thus, allowing meaningful comparisons to the bistatic synchronisation measurements at longer baselines. The above calibrations were carried out using the low-cost DMTD testbench described in Appendix C. First, the correct operation of the internal M12+ GPS receivers was verified by comparing the measured performance to the published literature. Then, the relative GPSDO hold-over performance was measured. The optimal PLL time constant was estimated at 1000 s by comparing the STALO to the GPS PPS reference during hold-over. The measured closed-loop performance confirmed the correct estimation of the optimal PLL constant. At zero-baselines, the time error can be calibrated to within 1 ns. However, this is expected to deteriorate by tens of nanoseconds across longer baselines when using auto-surveyed antenna positions. The time drifted by ±5.5 ns and ±3 ns over eight hours for $\tau_{PLL} = 150$ s and $\tau_{PLL} = 1000$ s, respectively. As expected, the long-term frequency offset and drift reached atomic levels of 10^{-13} and 10^{-14}, respectively. However, the instantaneous frequency accuracy was limited to $\pm2.6 \times 10^{-10}$ and $\pm1.3 \times 10^{-10}$ at $\tau_{PLL} = 150$ s and $\tau_{PLL} = 1000$ s, respectively. These frequency accuracies translate to maximum phase gradients of $224.6\,°/\mathrm{s}$ and $112.32\,°/\mathrm{s}$ which limits the bistatic coherent pulse integration losses to 6.5 dB and 1.4 dB, respectively. Populating the lower noise output buffer amplifiers is expected to lower the phase noise by about 15 dBc/Hz and 20 dBc/Hz which should improve the instantaneous frequency accuracy. However, this is left to future work. Refer to Section 5.6 for a full summary of the measured results.

In Chapter 6, a tristatic synchronisation experiment was set up in Simon's Bay, South Africa, where the individual NetRAD nodes were synchronised using the well-calibrated UCT GPSDOs. This chapter reported on the bistatic time, phase, frequency, and Doppler performance of the GPS-synchronised NetRAD. LOS phase compensation, where the direct breakthrough phase was used as the reference, improved the bistatic performance to near monostatic levels. Finally, the models developed in Section 3.4 for non-coherent and coherent pulse integration in the presence of a constant bistatic frequency offset were verified using real data.

- Firstly, NetRAD a coherent tri-node pulsed-Doppler networked radar was converted from a cabled system to an entirely wireless system. This required collaboration between UCL and UCT and a team of students, academics and industry experts. The author was responsible for integrating the UCT GPSDOs within the system. This resulted in the first published example of pulsed-Doppler phase synchronisation using GPS [27, 34]. The GPSDO EP mechanism enabled accurate and reliable time synchronisation. The quick-locking adaptive PLL enables the practical use of long PLL time constants. [35] The flexibility afforded by baselines up to a few kilometres enabled several bistatic sea clutter and small sea vessel trials along the Southern coasts of England and South Africa [100–105].

- Then, a tristatic experiment was set up in Simon's Bay, South Africa to assess the GPS-synchronised bistatic pulsed-Doppler performance. The Roman Rock lighthouse was used as a large and known static target while there were unobstructed LOS between the transmitter and receivers. This experiment produced the first published assessment of pulsed-Doppler phase synchronisation using GPS-synchronisation [35]. It is further the first published assessment of the efficacy of LOS phase compensation when used in combination with low-cost GPSDOs [109].

However, this experiment relied on self-surveyed GPS positions instead of precise and independently surveyed coordinates. Thus, the achieved time precision could not be measured. It is recommended that future experiments have precisely measured baselines.

- The time accuracy was measured by using the time of arrival of the direct breakthrough. These measurements were done for $\tau_{PLL} = 150\,\text{s}$, and the results were consistent with the measured $\pm 5.5\,\text{ns}$ zero-range time drift about the mean for the UCT GPSDOs.

These measurements compared the peaks of overlapping breakthrough range profiles over a 20-minute period. Future measurements should include more measurements over a longer period.

- It was shown that the relative GPSDO PLL time error is a good measure of both the range error and Tx-Rx frequency offset. Moreover, the relative GPSDO PLL time error superimposes near perfectly onto the RTI plot.

This suggests that the relative PLL time error could be used to correct range or frequency errors incurred during open-loop operation. However, further investigation is required.

- The phase, frequency, and Doppler performance of the monostatic and bistatic Roman Rock returns were compared. The bistatic phase measurements were consistent with the zero-baseline calibrations. However, as expected, the bistatic phase was significantly noisier than that of the monostatic.

It is unfortunate that the fixed PRF of 1 kHz along with the available recording memory limited the above experiments to 130 seconds each. Future measurements with more and longer recordings are highly desired. The fractional carrier frequency noise seemed to be dominated by noise originating from within NetRAD's 2.4 GHz LO synthesiser. This highlights the importance of matching the radar subsystem performance to that of the STALO. It is further desirable to compare the bistatic IF phase noise to that of the monostatic such that the relationship between the IF phase noise cancellation and SCV could be measured. However, due to the lack of opportunity and equipment, the IF phase noise was never measured. Instead, the aliased Doppler phase noise was plotted but is of somewhat limited use. The bistatic SCV was measured to be 30 dB worse than that of the monostatic. The bistatic range-Doppler plots were noisier with more pronounced 'tramlines' caused by the uncancelled spurious frequencies. However, in the absence of large targets, a relatively small speedboat produced reasonable Doppler performance.

- LOS phase compensation was applied to the measured bistatic Roman Rock phase. The recorded direct breakthrough was used as the phase reference. The phase compensated bistatic results were compared to both the bistatic and monostatic phases. Moreover, LOS phase compensation restored the bistatic phase, frequency, and Doppler performance to near monostatic levels. [109]

Thus, LOS phase compensation can be used in combination with low-cost GPSDOs to produce high-performance and near monostatic pulsed-Doppler performance. This method is both cost and computation efficient and could be performed in real-time in future implementations. However, GPSDO pre-synchronisation is still required to synchronise the Tx-Rx PRF and triggering. Excellent phase compensation was achieved across the short baselines of 2.3 km that had clear LOS. Future experiments should include longer baselines containing more clutter and multipath inducing elements.

- Finally, the effect of a constant bistatic frequency offset on both non-coherent and coherent integration was studied using real bistatic data. In both cases, the real data compared well with the theory of Section 3.4. A larger $1/f$ integrated noise component appeared to be advantageous under certain conditions for the coherent case.

However, more data and further analysis are required. In future research, range migration compensation techniques could be applied in an attempt to mitigate Tx-Rx frequency offsets and drifts.

At the time of the above bistatic recordings, the quick-locking adaptive PLL was not yet completed. Thus, these measurements were made using the suboptimal GPSDO $\tau_{PLL} = 150\,\mathrm{s}$. Nonetheless, these measurements compared well to the zero-range calibrations at $\tau_{PLL} = 150\,\mathrm{s}$. Future measurements may include measurements at the optimal $\tau_{PLL} = 1000\,\mathrm{s}$. The quick-locking adaptive PLL has since been successfully implemented in the newer NeXtRAD tristatic system [33] enabling measurements at optimal synchronisation.

Appendix A reviewed some basic concepts of frequency stability used throughout this text. These concepts included a systematic oscillator model, time- and frequency-domain frequency stability measures, and the DMTD technique used to make high-resolution frequency stability measurement. The oscillator model was used in Chapter 3 to model bistatic synchronisation behaviour. The frequency stability measures are used throughout the text to compare and characterise oscillator performance. Finally, the DMTD technique was used in Chapter 5 to calibrate the UCT GPSDOs.

Appendix B contains the datasheets of the Oscilloquartz OCXO 8788 used as STALO to the UCT GPSDO. The manufacturer has since stopped manufacturing OCXOs, and the datasheets may be difficult to find in the future.

Appendix C detailed the design and calibration of the low-cost UCT DMTD system. This DMTD system was developed with the purpose of performing repeatable and reliable high-resolution phase and frequency calibrations. For this reason, each subsystem is meticulously documented. The DMTD noise floor is also carefully calibrated to ensure the validity of subsequent oscillator measurements. During a 60-hour noise floor measurement, the phase drifted by only 30 ps with a short-term peak-to-peak noise of 10.8 ps with a standard deviation of 0.73 ps. The frequency offset and drift were , and the frequency noise had ± 5 MAD boundaries of $\pm 6.19 \times 10^{-14}$. The UCT DMTD has a frequency stability noise floor of $\sigma_y(\tau = 1) = 1.65 \times 10^{-13}$ which drops below 10^{-14} at 200 seconds. This noise floor is well below the current and near-future requirements of the UCT RF laboratory.

- Publications related to RF design for frequency stability that are not well-known to the radar community were gathered. Recommendations were made on how to apply these frequency stability techniques to future high-fidelity coherent bistatic radar. (see Section C.3)

Appendix A

Frequency Stability

This appendix reviews some basic concepts of frequency stability. First, a systematic oscillator model from [87, 88] is introduced describing an oscillator's output in terms of both systematic and random effects. This model is used in Chapter 3 to predict pulsed-Doppler radar behaviour as a function of oscillator performance. Then, various measures of frequency stability is discussed. These measures are used to calibrate the UCT GPSDOs in Chapter 5 as well as to help quantify the performance of the GPSDO synchronised NetRAD in Chapter 6. Finally, a method to perform high-resolution frequency stability measurements in the time domain is reviewed. This method is called the DMTD technique and can be used to measure time, time fluctuations, frequency and frequency fluctuations of two or more equal frequency stable oscillators [110]. Moreover, the DMTD method was used to calibrate the UCT GPSDOs using the low-cost DMTD system described in Appendix C.

The field of frequency stability is a mature discipline. The reader may consult the following sources for a more in-depth review of the field. The NIST technical note 1337 [111] contains a collection of papers on the characterisation of oscillators and is the authoritative reference on the subject. The first four papers in this collection [87, 88, 110, 112] serve as introductory material to the field of oscillator characterisation. The standard accepted terminology relating to time and frequency characterisation is defined in [113]. More recently, Riley [114] produced a comprehensive summary of time and frequency characterisation methods.

A.1 Systematic Oscillator Model

In this section a systematic oscillator model for phase synchronisation is reproduced from [87, 88]. This model expresses an oscillator's signal in terms of initial time offset, initial frequency offset, a constant frequency drift, and random phase fluctuations. The model can be used to predict the time and frequency error as a function of future time [114] and is used in Chapter 3 to model the effect of imperfect bistatic synchronisation.

An oscillator's signal can be represented by

$$V(t) = [V_o + \epsilon(t)] \sin[\omega_o t + \phi(t)], \tag{A.1}$$

where $V(t)$ is the instantaneous output voltage, V_o is the nominal output amplitude, and ω_o is the nominal output frequency. The time dependent voltage and phase deviations are represented by $\epsilon(t)$ and $\phi(t)$, respectively. The signal amplitude $[V_o + \epsilon(t)]$ is assumed to be unity to model oscillator phase synchronisation. Frequency is the derivative of phase [87], and the instantaneous frequency is

$$v(t) = v_o + \frac{d\phi(t)}{2\pi dt}, \qquad [\text{Hz}] \quad (A.2)$$

where $\omega_o = 2\pi v_o$ [rad/s]. The instantaneous frequency can be normalised to the nominal output frequency and Walls et al. [88] defines the fractional frequency deviation from the nominal as

$$y(t) = \frac{v(t) - v_o}{vo} = \frac{d\phi(t)}{2\pi dt}. \qquad [\,] \quad (A.3)$$

From [87], the integral of fractional frequency deviation gives time deviation

$$x(t) = \int_0^t y(t')dt', \qquad [\text{s}] \quad (A.4)$$

where

$$x(t) = \frac{\phi(t)}{2\pi v_o}. \qquad [\text{s}] \quad (A.5)$$

From [84, 87, 114], an oscillator's time deviation from the ideal can be modelled as

$$x_\epsilon(t) = x_o + y_o t + \Delta T K_T t + \frac{1}{2} D t^2 + \sigma_x(t) \qquad \text{[s]} \quad \text{(A.6)}$$

where x_o is the initial time offset, y_o is the initial fractional frequency offset, ΔT is the temperature offset, K_T is the temperature stability coefficient, D represent a constant frequency drift (oscillator ageing), and $\sigma_x(t)$ the time dependent time deviation due to the random phase fluctuations. From (A.6) in (A.3) one may also write

$$y_\epsilon(t) = y_o + Dt + \Delta T K_T + \sigma_y(t), \qquad \text{[ ]} \quad \text{(A.7)}$$

where $\sigma_y(t)$ is the random time dependent frequency deviation due to the random frequency fluctuations. Thus, $x_\epsilon(t)$ and $y_\epsilon(t)$ represents the cumulative time and frequency errors from the ideal after an arbitrary time period t, respectively.

Note that some oscillators may exhibit significant frequency modulation or non-constant frequency drift [87]. The models in (A.6) and (A.7) are sufficient to help with the analysis of radar synchronisation in the presence of ideal systematic effects, and random phenomena. However, many real-world scenarios would require more complex oscillator models.

A.2 Frequency Domain Stability Measures

The one-sided spectral density of fractional frequency, $S_y(f)$, is generally accepted as the preferred measure of frequency stability in the frequency domain [113]. However, the spectral density of phase, $S_\phi(f)$, is often more convenient [115]. From [113] the two measures are related by

$$S_y(f) = \left(\frac{f}{v_o}\right)^2 S_\phi(f). \qquad \text{[Hz}^{-1}\text{]} \quad \text{(A.8)}$$

It is customary to plot $S_\phi(f)$ on a log-log scale in units of dBc/Hz [32]. From [115] the oscillator spectral densities can be modelled using power-law curves such that

$$S_y(f) = \begin{cases} \sum_{\alpha=-2}^{+2} h_\alpha f^\alpha, & 0 \leq f \leq f_h \\ 0, & f > f_h, \end{cases} \qquad \text{[Hz}^{-1}\text{]} \quad \text{(A.9)}$$

Table A.1: Slope Characteristics for the Different Noise Types using Various Stability Measures [113]

	S_y	S_ϕ	$\sigma_y^2(\tau)$	$\sigma_y(\tau)$	$\mathrm{Mod}_{\sigma_y}^2(\tau)$
	α	β	μ	$\mu/2$	μ'
Random Walk FM	-2	-4	1	1/2	1
Flicker FM	-1	-3	0	0	0
White FM	0	-2	-1	-1/2	-1
Flicker PM	1	-1	-2	-1	-2
White PM	2	-0	-2	-1	-3

where the exponent α is indicative of the specific noise type, the constant h_α is representative of the noise power, and f_h is the model spectral density cutoff frequency. Moreover, the value of α takes on integer values between -2 and +2 such that the different noise types are each represented by a constant slope on a log-log plot. The phase spectral density $S_\phi(f)$ can be modelled in a similar fashion. However, the power-law slopes are different after conversion from $S_y(f)$ using (A.8). It is customary to represent the slopes of $S_\phi(f)$ using $\beta = \alpha - 2$. Refer to Table A.1 to see how the power-law slopes are related to the different noise types. The noise types usually found in oscillator spectral densities are Random Walk FM, Flicker noise and White noise. Flicker and white noise originate from the electronic components and can both frequency modulate (FM) or phase modulate (PM) the signal depending on the mechanism. Random Walk FM is often related to changes in the oscillators environment such as temperature, humidity and vibration. Refer to Rutman [115] for a more in-depth discussion on the origin of the different noise types.

It is also of interest to note that equivalent expressions for phase noise can be derived from narrowband FM theory. See Goldman [32] for more details.

A.3   Time Domain Stability Measures

Variances are used to characterise time domain stability. The standard variance is a poor measure of frequency stability because for certain noise types it does not converge. [114] Thus, other variances were developed over the years which serve as robust and converging statistical measures of frequency stability. In this section, the Allan, Overlapping Allan, Modified Allan and Allan Time Variances is briefly introduced from [111, 114].

Time domain frequency stability can be characterised by measured time interval data using the classical Allan Variance, AVAR, which is calculated as

$$\sigma_y^2(\tau) = \frac{1}{2(N-2)\tau^2} \sum_{i=1}^{N-2} [x_{i+2} - 2x_{i+1} + x_i]^2, \tag{A.10}$$

where x_i is the ith time interval measurement in units of seconds, τ is the measurement interval, and N is the total number of samples in the data set.

However, the classical AVAR has relatively poor confidence and make inefficient use of the available data. Thus, the Overlapping Allan Variance, OVAR, was developed to calculate the AVAR by using all the possible overlapping samples for each averaging period. Thus, making more efficient use of the data and significantly improving the confidence level. The OVAR is the most often used measure of frequency stability today [114]. The Overlapping Allan Variance is calculated as

$$\sigma_y^2(\tau) = \frac{1}{2(N-2m)\tau^2} \sum_{i=1}^{N-2m} [x_{i+2m} - 2x_{i+m} + x_i]^2, \tag{A.11}$$

where $\tau = m\tau_o$, τ_o is measurement interval, and m is the averaging factor. Thus, the fundamental measurement interval measurement, τ_o, is altered by the averaging factor m for all combinations.

The slopes of AVAR and OVAR plots on log-log scales can be used to identify the various noise types. Similar to the log-log slopes of the frequency domain power-law spectra. However, AVAR and OVAR cannot unambiguously differentiate between Flicker PM and White PM noise. Refer to Table A.1 where the power-law slopes of the various noise types are compared for the various statistical measures. The slope of the AVAR log-log power-law is denoted β. However, the Modified Allan Variance, MVAR, was developed to differentiate between Flicker PM and White PM noise unambiguously. The slope of the AVAR log-log power-law is denoted β'.

The MVAR is calculated as

$$\text{Mod}^2_{\sigma_y}(\tau) = \frac{1}{2m^2\tau^2(N - 3m + 1)} \sum_{j=1}^{N-3m+1} \sum_{i=j}^{j+m-1} [x_{i+2m} - 2x_{i+m} + x_i]^2. \qquad (A.12)$$

The Time Allan Variance, TVAR, is derived from the MVAR to serve as a measure of time stability, and is calculated as

$$\sigma_x^2(\tau) = (\frac{\tau^2}{3})\text{Mod}^2_{\sigma_y}(\tau). \qquad (A.13)$$

The TVAR is used as a measure of the time stability of time distribution networks [114].

It is customary to plot the above variances as deviations by computing the sqaure root of the variance. The deviations of the AVAR, OVAR, MVAR and TVAR are denoted as ADEV, ODEV, MDEV and TDEV, respectively.

The time and frequency domain measures of frequency stability are equivalent and it is possible to convert between the two domains. The AVAR can be calculated from single sideband spectral density of frequency stability, $S_y(f)$, using the following equation

$$\sigma_y^2(\tau) = 2 \int_0^{f_h} S_y(f)\frac{\sin^4(\pi\tau f)}{(\pi\tau f)^2}. \qquad (A.14)$$

It is also possible to convert from the time domain to the frequency domain using the tables given in [113].

A.4 The Dual-Mixer Time Difference Technique

Direct frequency stability measurement using a TIC lacks the resolution to measure stable oscillators. However, various methods exist to make high-resolution frequency stability measurements. The DMTD technique is one such technique and increases measurement resolution by utilising the heterodyne effect. Moreover, the resolution can be increased by a factor of 10^7 or more when compared to a direct measurement. The DMTD is well understood and has been in use for decades [116–123]. It enables high-resolution measurements of time, time fluctuations, frequency and frequency fluctuations of two equal frequency oscillators [110]. Furthermore, the absence of

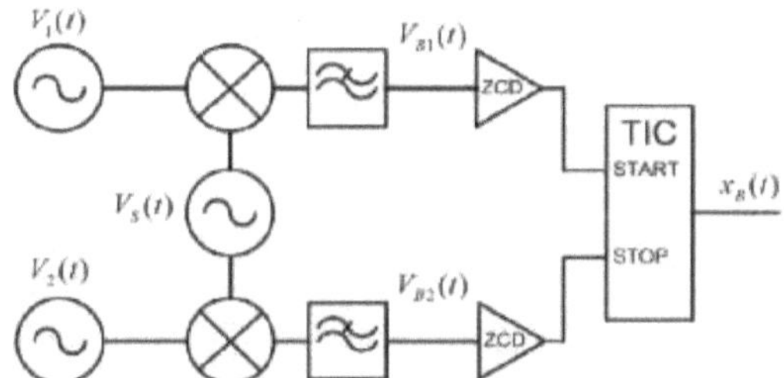

Figure A.1: Block Diagram of a Classical DMTD System.

measurement dead time is beneficial when calculating the Allan deviation [112]. The ability to measure time directly is of particular interest to bistatic radar research.

This section provides a brief explanation of the DMTD technique. Consult [110, 112, 124] for a more in-depth discussion. The relevant equations are reproduced below to aid in the discussion.

A classical DMTD system is depicted in Figure A.1. Oscillators 1 and 2 are compared where S is the reference oscillator. Moreover, all three oscillators are assumed to be at the same nominal frequency with the reference oscillator offset by a few Hertz. Oscillators 1 and 2 are simultaneously mixed with the reference oscillator using two balanced mixers. After IF lowpass filtering, two beat notes are produced with a nominal frequency equal to that of the reference offset. The time difference between the zero-crossings of the two beat notes is equal to the heterodyne scaled time difference between oscillators 1 and 2. A time interval counter records this time difference.

From [112][1] the three oscillator voltages could be represented by

$$V_1(t) = [V_{10} + \epsilon_1(t)] \sin[2\pi v_{10}t + \phi_1(t) + \phi_{10}] \tag{A.15}$$

$$V_2(t) = [V_{20} + \epsilon_2(t)] \sin[2\pi v_{20}t + \phi_2(t) + \phi_{20}] \tag{A.16}$$

$$V_S(t) = [V_{S0} + \epsilon_S(t)] \cos[2\pi v_{S0}t + \phi_S(t) + \phi_{S0}] \tag{A.17}$$

where V is the nominal amplitude, $\epsilon(t)$ the amplitude fluctuation, v the nominal

---

[1] $\phi_{10},\phi_{20},\phi_{S0}$ were added to include the initial phase offset.

frequency, $\phi(t)$ the phase fluctuation, and ϕ the initial phase offset. Also from [112] the low-pass filtered beat frequencies at the mixer's IF can be written as

$$V_{B1}(t) = V_{B10} \cos[2\pi(v_{10} - v_{S0})t + \phi_1(t) - \phi_S(t) + \phi_{10} - \phi_{S0}] \qquad \text{(A.18)}$$

$$V_{B2}(t) = V_{B20} \cos[2\pi(v_{20} - v_{S0})t + \phi_2(t) - \phi_S(t) + \phi_{20} - \phi_{S0}] \qquad \text{(A.19)}$$

The phase difference is measured by comparing the beat note zero-crossings. Zero-crossing detectors amplify the slopes of the beat note zero-crossings. Thus, amplitude variations are assumed to be negligible. This phase difference can be written as

$$\Delta\phi(t) = \Delta\phi_{12}(t) + \Delta\phi_{12} \qquad \text{[rad]} \quad \text{(A.20)}$$

where $\Delta\phi_{12}(t) = \phi_1(t) - \phi_2(t)$ is the time dependant phase fluctuation, and $\Delta\phi_{12} = \phi_{10} - \phi_{20}$ is the initial phase offset, between oscillators 1 and 2. The phase terms of the reference oscillator are common to both and cancel.

The nominal beat frequency can be approximated as

$$v_B \approx |v_{10} - v_{S0}| \approx |v_{20} - v_{S0}| \qquad \text{[Hz]} \quad \text{(A.21)}$$

when one assumes that $|v_{10} - v_{20}| \ll 0$. The DMTD technique requires that the oscillators under test, as well as, the reference oscillator, are nominally at the same frequencies. The heterodyne factor [124] is given by $R = v_B/v_o$ where v_o is the nominal oscillator frequency. Phase in units of time [112] is given by $x(t) = \phi(t)/2\pi v_o$. Now (A.20) can be rewritten as

$$x(t) = x_B(t)/R \qquad \text{[s]} \quad \text{(A.22)}$$

to give the time difference, $x(t)$, between oscillators 1 and 2, where $x_B(t)$ is the TIC measured time difference.

Thus, it can be seen that the DMTD technique improves the measurement sensitivity by a factor R. For $v_B = 10\,\text{Hz}$ and $v_o = 10\,\text{MHz}$ the sensitivity is improved by $R = 10^6$.

Appendix B

Oscilloquartz OCXO 8788 Specification

Oscilloquartz has since stopped manufacturing OCXOs, and data sheets may be difficult to find in the future. Hence, the datasheet for the Oscilloquartz 8788 OCXO is included in this appendix. Additionally, the phase noise, temperature stability, and frequency drift as measured by Oscilloquartz are also included.[1]

---

[1]Many thanks to Yves Scwabb who was so thoughtful to provide these measurements.

Technical
Specifications

OCXO 8788/8789

Oven Controlled Crystal Oscillator

Standard / Option	Standard	Option
Crystal Oscillator	SC-cut	
Standard frequencies	10 MHz / 5 MHz	Consult factory
Operating temperature range	A: -20°C to $+70^\circ$C B: 0°C to $+70^\circ$C C: 0°C to $+60^\circ$C	D: -10°C to $+70^\circ$C E: -40°C to $+70^\circ$C
Frequency stability (Δ f/f)		
Long term stability (aging after 30 days of continuous operation)	Standard: 5×10^{-10}/day 7×10^{-8}/year	G: 2×10^{-10}/day 3×10^{-8}/year
Setting @ 25°C VC max / 2	$< \pm 2\times10^{-7}$	
Over temperature range (Y)	Std : $< 2\times10^{-8}$ peak to peak	over consult factory
Versus supply voltage changes (Vcc $\pm$ 5%)	$< \pm 2\times10^{-10}$	
Versus load changes (50Ω $\pm$ 10%)	$< \pm 2\times10^{-10}$	
Short term stability σ (τ) @ is Allan variance	$< 1\times10^{-12}$	
Electronic frequency control : Z > 10 kΩ	$>\pm$ 0.8 ppm (0 to 10 Volts) / Linearity<10% / Positive slope	
Power Supply (P)		
Input voltage range (DC)	+12 Volts $\pm$ 10% / 24V $\pm$ 10%	
Power consumption (@Vcc = 12V)	< 8W during warm up / < 2.5W after warm-up at +25°C	
Environment (Not operating)		
Storage temperature	-40°C to $+125^\circ$C	
Vibration	IEC 68-2-6 Test Fc : 10 Hz—500 Hz, 10g	
Shock	IEC 68-2-27 : Half-sine 50g, 11ms	
Size (L x W x H)	8788: 51.1 x 41.1 x 19mm	8789: 50.8 x 50.8 x 19mm (2"x2"x0.75")
Weight	$\geq$ 80g	
Outline and electrical connections	See drawing	
Outputs Characteristics (Z)	S	
Wave form	Sine	
Level (Tol.) / Impedance	8dBm $\pm$ 1 dBm / 50 Ω	
Phase noise	See drawing	
Harmonics	< -25 dBc	
Spurious in the frequency range up to 1 MHz	< -75 dBc	

Oscilloquartz SA reserves the right to change all specifications contained herein at any time without prior notice.

Ordering Information

8789 — B SG - 5.000 MHz

Model
+12Vdc

Operating temperature range
A; B; C; D; E

Aging Options
- ; G

Nominal frequency output
5.000 MHz

Output signal
S: Sine wave

Edition 3/Oct. 2004/OBS

A COMPANY OF THE SWATCH GROUP

www.oscilloquartz.com

OSCILLOQUARTZ
SWATCH GROUP ELECTRONIC SYSTEMS

Figure B.1: Oscilloquartz OCXO 8788 Specification

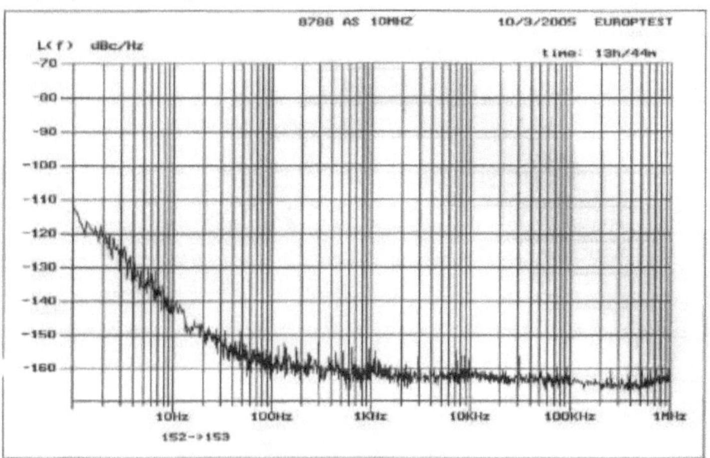

Figure B.2: Oscilloquartz OCXO 8788 Phase Noise

OSCILLOQUARTZ S.A.

Rue des brévards 16, CH-2002 Neuchâtel 2, SWITZERLAND

Phone: (+41) 32 722 55 55   Fax: (+41) 32 722 55 56

Email: osa@oscilloquartz.com   Web site: www.oscilloquartz.com

Measurements results

Measurements date:	9/3/2005 @ 13:16:36	**OF Nr:** 18777
Serial Nr:	155	**Option(s):**
Oscillator type:	8788	(B) 2.47e-09 : YES
Oscillator frequency:	10000.000 KHz	(A) 3.97e-09 : YES
Oscillator P/N:	945.878.810.00	

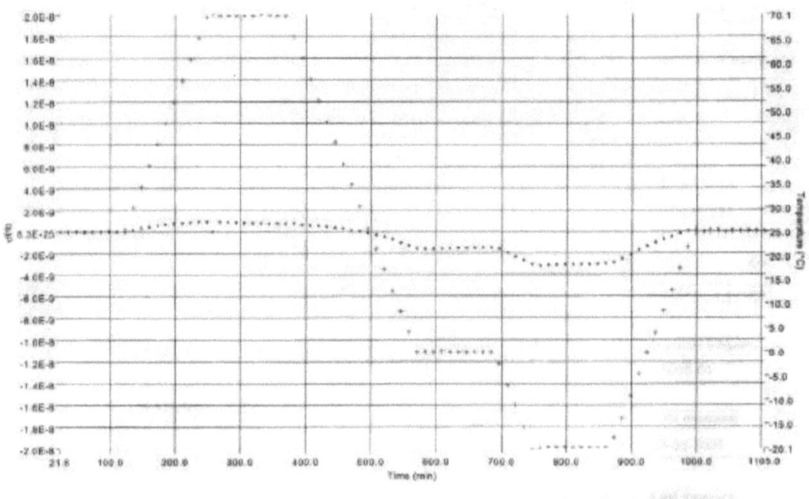

Figure B.3: Oscilloquartz OCXO 8788 Temperature Stability

191

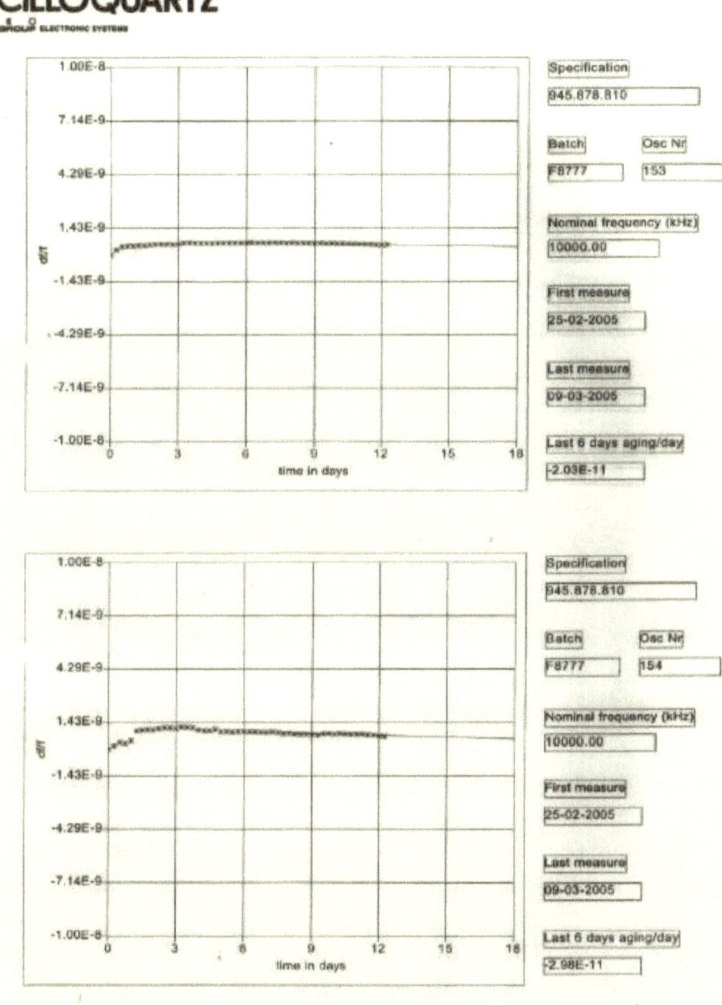

Figure B.4: Oscilloquartz OCXO 8788 Frequency Drift

Appendix C

The UCT Dual-Mixer Time Difference System

This appendix describes the design and calibration of the low-cost UCT DMTD system. High-resolution measurement of time and frequency stability is an essential part of GPSDO design, verification and re-calibration. The purpose of the UCT DMTD is to calibrate the UCT GSPDOs and other laboratory standards. At the time, UCT did not have access to specialist equipment to do high-resolution frequency stability calibrations. Moreover, such measurements can often last weeks or months making outside laboratories or loan equipment impractical. An in-house DMTD added significant capability in both equipment and expertise to the research group. Valuable insight is gained into frequency stability design some of which is directly applicable to coherent bistatic radar design. To date, the UCT DMTD has been used to calibrate the UCT GPSDOs (see Chapter 5) and a CERN WR fibre synchronisation system [41, 42]. It is currently being used in the development of NeXtRAD.

The DMTD frequency stability measurement technique is well understood and has been in use for decades [116–123, 125]. Direct measurement using a TIC often lacks the resolution to measure stable oscillators. However, this resolution can be increased by a factor of 10^7 or more with the DMTD technique which uses the heterodyne effect. This makes the very precise measurements of time (phase), frequency and frequency stability possible. Refer to Section A.4 for a review of the DMTD technique.

The UCT DMTD is rather versatile and is based on a design by Riley [117].

However, a different zero-crossing detector (ZCD) strategy was used, and similar to [119], an external LO and TICs are used. The use of standard external laboratory LOs and TICs simplifies the design, shortened the design time, and allows more flexibility. An Agilent E4400B signal generator was used as LO along with two HP53131A TICs. The system caters for oscillator frequencies of 1 MHz to 100 MHz and beat frequencies of 1 Hz to 100 Hz. Four channels make it possible to compare up to four oscillators simultaneously. Similar to Riley's design, the system comprises of two identical DMTD sections making ultra-low noise cross-correlation measurements possible. However, the cross-correlation capability remains untested to date. Three-cornered hat measurements are also possible but require a third TIC.

The system achieved an ODEV noise floor of 1.65×10^{-13} at 10 MHz and a 10 Hz beat with a peak time drift of 30 ps over a 60 hour period. This noise floor is well below our current and near future requirements. However, the noise floor may be lowered further by swapping the Agilent E4400B with a more stable LO. Cross-correlation is expected to lower the noise floor by an order of magnitude.

Design for frequency stability is an established field, but not well known to the radar community. Therefore, this appendix gives a detailed account of the DMTD design. The various subsystems are explained which includes the LO distribution circuitry, the mixer isolation amplifiers, the mixer and IF filtering, the zero-crossing detector, the power supply, and data recording. The VSWR stabilising techniques that are applicable to high-performance coherent bistatic radar are highlighted. Finally, the two HP53131A TICs are calibrated, and the UCT DMTD noise floor is measured.

C.1 Design of the UCT DMTD System

This section details the design of the UCT DMTD system. The system consists of four channels forming two identical DMTD sections. Refer to the diagram in Figure C.1 and photos in Figure C.2. Each channel consists of a double balanced mixer, isolation amplifiers at the mixer LO and RF ports, an IF low-pass filter and a ZCD. An external LO is distributed to the mixer LO ports by way of a 1:4 splitter. The ZCD outputs are directly available to the user, and external TICs are required. Three-cornered hat[1] measurements are possible using three TICs [117]. The system

---

[1] A three-cornered hat calibration measures the relative frequency stability of three oscillators simultaneously. The absolute frequency stability of each oscillator is then estimated through what

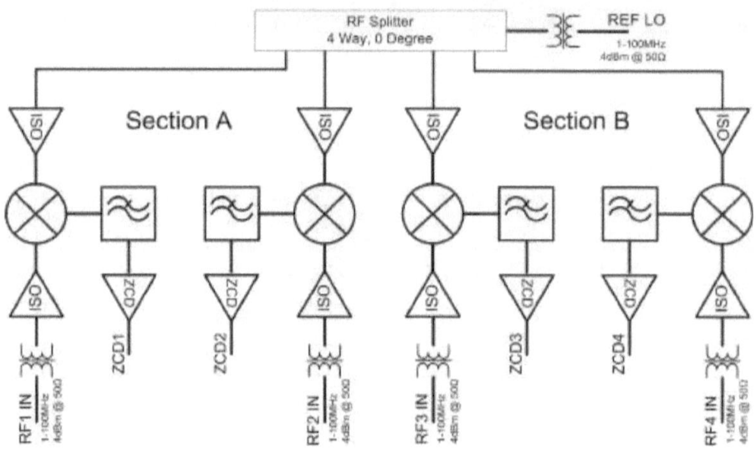

Figure C.1: Block Diagram of the UCT DMTD System.

caters for oscillator frequencies of 1 MHz to 100 MHz and beat frequencies of 1 Hz to 100 Hz. Up to four oscillators can be compared simultaneously. Frequency references are often available with 6 dBm to 13 dBm @ 50 Ω outputs. Therefore, the LO and RF inputs are set at 4 dBm allowing for an external attenuation pad to improve the cable VSWR, which in turn improves frequency stability [126]. This also conveniently allows the RF inputs to be fed directly from the two 1:2 splitters included on the front panel. The system uses SMA RF connectors to reduce phase instability due to mechanical effects. The UCT lab also uses SMA connectors extensively. However, the flexible RG-178 internal cabling increases the vibration sensitivity. All inputs are transformer isolated to minimise ground loop effects. Metal film chip resistors and multi-layered ceramic chip capacitors are used throughout the design. NPO and C0G capacitor dielectrics were used where possible to reduce the filter temperature coefficients. An Agilent E4400B signal generator was used as the LO along with two HP53131A TICs.

The following sections detail the individual subsystems. These subsystems include the LO clock distribution, the isolation amplifier, the mixer and IF filtering, and the ZCDs. The purpose is not only to document the DMTD design but also to point out the high stability design techniques and references applicable to high-

---

is called the separation of variances [110].

performance coherent bistatic systems. Refer to the schematics in Figures C.8 to C.10.

C.1.1   Local Oscillator

The 4 dBm LO input is transformer isolated and split 1:4 using a Mini-Circuits AD4PS-1+ splitter. This splitter was chosen on its excellent VSWR < 1.12 and good isolation > 30 dB. The DMTD technique is based on the assumption that the LO's noise contribution cancel. Moreover, Sze-Ming [127] shows that as the time difference between two beat notes approaches zero the noise contribution of the LO also approaches zero. In standard DMTD measurements, the beat note offset is nulled by manipulating the RF input phase. However, when performing cross-correlation measurements, the beat note offsets cannot be nulled simultaneously in both DMTD sections. Therefore, the splitter outputs are routed to their respective mixer inputs via equidistant PCB tracks. However, there remains some concern over the relatively high splitter phase unbalance of 2° to 5° at 10 MHz to 100 MHz. It would have been ideal to use the PCB track length to also compensate for the mixer phase unbalance and isolation amplifier part-to-part skew.

C.1.2   Isolation Amplifiers

Isolation amplifier phase noise and harmonic distortion has a direct influence on the system's frequency stability [126, 128]. There are discrete low-phase noise and high isolation amplifier circuits published in the literature, see [129]. However, broadband low-noise opamps were used instead. Opamps are easier to design with a lower component count simplifying the construction. Inconveniently, opamp phase noise and reverse isolation are not usually specified. Nonetheless, some 'time-nut' members [130] reported excellent phase noise and reverse isolation performance for some low-noise opamps. The LMH6609 opamp was chosen for mixer isolation due to its proven performance in the Riley design [117]. It is an ultra-wideband 900 MHz voltage feedback opamp with balanced and symmetrical inputs. It has high load drive capability with low harmonic distortion. It also has a low $1/f$ voltage noise knee of $4 \, \text{nV}/\sqrt{\text{Hz}}$ @ 2 kHz.

Step-up RF transformers (T1; T2) are used to produce a $2 \, V_{pp}$ signal at the input of each LMH6609. The high input impedance of the non-inverting opamp helps to $50 \, \Omega$ match the transformer inputs (R24; R33). Using a step-up transformer is

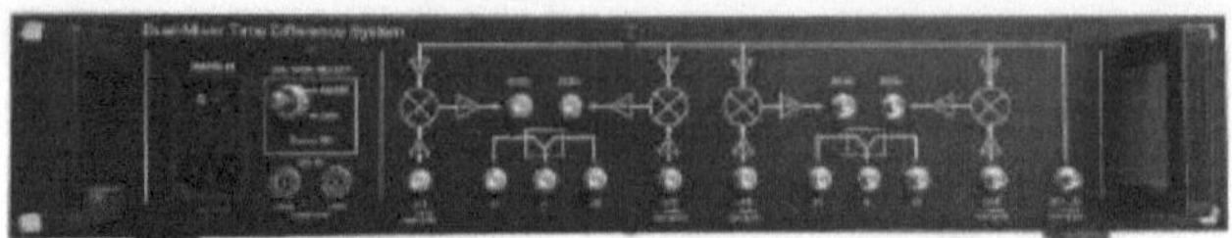

(a) Enclosure Front View

(b) Enclosure Top View

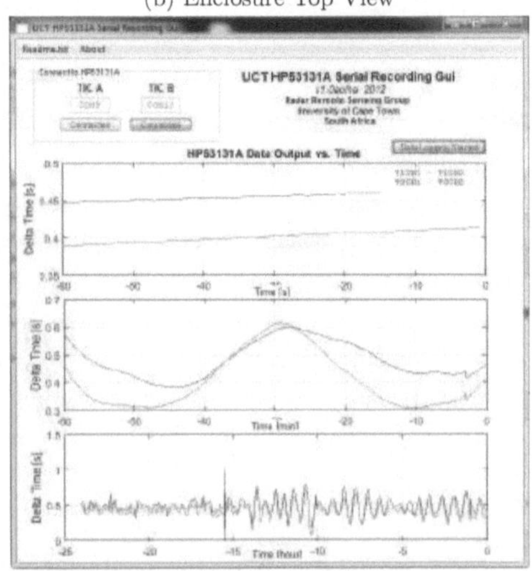

(c) DMTD Recording GUI

Figure C.2: The UCT Dual-Mixer Time Difference System

convenient because the signal and noise power increase by the same ratio keeping the spectral density of phase unchanged [129].

The opamp gains are set to deliver 10 dBm to the mixer's RF and LO ports. Low-value feedback resistors were used to limit their noise contribution. DC-block capacitors (C35; C48), suggested by Griffiths [131], eliminate the DC currents through the feedback resistors to ground and prevents amplification of the DC signal.

The opamp's $\pm 5\,$V power rails are heavily decoupled and have series $10\,\Omega$ resistors (R20; R26; R28; R34) for additional filtering above 723 Hz. The opamp has excellent 3rd harmonic distortion ($< 80\,$dBc) for the 10 dBm output below 10 MHz. However, the distortion deteriorates rapidly ($< 40\,$dBc) at 50 MHz. Only odd harmonics introduce beat note time errors since even harmonics cancel in a double balanced mixer. Harmonic time errors fall off as $1/n$ where n is the harmonic number [128].

C.1.3 Mixer and IF Filtering

The Level 7 Mini-Circuits SBL-1+ double balanced mixer was chosen for its high isolation and excellent conversion loss of approximately 5.6 dB between 10 MHz and 100 MHz. The performance data show that the isolation exceeds 60 dB below 20 MHz.

The isolation amplifiers drive the LO and RF ports at 10 dBm saturating the mixer. Extra resistance (R25; R30) was added at the LO and RF inputs as suggested by Nelson and Walls [132]. These resistors improve VSWR and make the system less susceptible to power and mixer temperature variations. Nelson and Walls [132] further recommended 3 dB to 6 dB pads in addition to the extra resistance to further stabilise the VSWR against power and temperature variation. However, this would have required a higher isolation amplifier output increasing harmonic distortion.

The mixer IF is terminated capacitively and in high resistance as recommended by Griffiths [131]. Terminating the mixer like this, along with saturating the LO and RF inputs, minimises noise by maximising the beat note's zero-crossing slope. A trapezoidal beat note with an amplitude of $0.7\,$V$_{\text{pp}}$ and zero-crossing slope of 36V/s at 10 Hz was measured. For the IF lowpass filter, the approach in [119] was followed using a passive inductor-capacitor (LC) stage followed by a passive resistor-capacitor (RC) stage. The LC-stage has a 3 dB bandwidth of 21 kHz. The magnetically shielded filter inductor has an self-resonant frequency (SFR) of 18 MHz. The RC-stage is intended to compensate for the high-frequency components not filtered by the LC-stage. However, the RC-stage was later removed because it had little effect and it forms an unwanted voltage divider with the ZCD input stage. The IF filter bandwidth was chosen such that the 1 Hz to 100 Hz region has a flat phase response to reduce temperature induced phase effects.

An aluminium plate thermally couples the four mixers. This is an attempt to improve the mixer temperature tracking. At 10 MHz mixers can have phase shift temperature coefficients of tens of ps/°C [132]. However, the distance between the mixers may be too large for this to be effective. A copper plate may improve thermal tracking.

C.1.4 Zero-Crossing Detector

The ZCD is slope amplifies the mixer output to minimise the TIC input noise caused by amplitude to phase conversion. The UCT DMTD has four-stage Collins style ZCDs. Collins [133] has shown that the noise contribution of a ZCD can be minimised by amplifying and low-pass filtering the input signal in several stages. At each stage, the amplification and filter bandwidth is increased. In our case, the mixer output has a measured zero-crossing slope of 36 V/s at 10 Hz. Total ZCD gain was set to approximately 10^6 resulting in an output slope of 36 V/µs and calculated RMS jitter of 4.84 ns. Note that the jitter calculation is only an estimate since it assumes identical noise contributions of $10\,\mathrm{nV}/\sqrt{\mathrm{Hz}}$ by all four stages. Refer to Table C.1 for the calculated and actual gain and bandwidth of each of the four stages. The calculated and actual values differ due to component availability. The actual ZCD performance may be improved by choosing better component values.

The first opamp stage is inverting and linear without any signal clipping. The mixer noise is likely to dominate, and it is recommended to increase the input and feedback resistance values (R5;R51) such that a higher resistance is presented to the mixer IF output. Resistor R51 protects the mixer from inadvertent fault currents [131]. This first stage is followed by an RC-highpass filter to remove the temperature dependent mixer DC offset. Each of the remaining stages consists of a non-inverting opamp followed by an RC-lowpass filter. The opamp outputs are clamped using diodes. Additional clamping diodes across the opamp feedback resistors prevent the opamp from saturating as suggested by Griffiths [131]. The LT1007, an 8 MHz ultra-low noise voltage feedback opamp was used for the first three stages. It has a wideband noise of $2.5\mathrm{nV}/\sqrt{\mathrm{Hz}}$ and a 1/f knee at 2 Hz. The gain of the fourth

Table C.1: Zero-Crossing Detector Gain and Bandwidth Settings

Stage	Calc. Gain	Calc. Bandwidth	Actual Gain	Actual Bandwidth
1	-5.36	49 Hz	-5.24	68 Hz
2	11.4	446 Hz	10.94	596 Hz
3	52.18	18.56 kHz	47	22 kHz
4	1080	16.4 MHz	906	16.37 MHz

and final stage is too high for a single opamp and was split across two higher gain-bandwidth AD829s. Both the LT1007 and AD829 have excellent common mode rejection ratio (CMMR) and power supply rejection ratio (PSSR).

Each opamp is heavily decoupled, and care was taken to keep the feedback resistance values low to limit their noise contributions. The passive RC-filter stages are expected to dominate temperature induced phase effects.

In hindsight, the ZCD gain was probably set too high, and the ZCD performance may be improved by lowering the gain by a factor of ten. Another observation is that the mixer IF filtering stage may be redundant. The capacitive-resistive loading at the IF output limits the mixer's frequency response, and the first stage ZCD filtering may well be adequate.

C.1.5 Power Supply Unit

The UCT DMTD can be powered off of mains or an external $\pm 12\,V_{DC}$ supply. The input voltage is selectable through the front panel. The mains power supply is a standard power transformer based linear power supply. There are X2 polypropylene filter capacitors (C9; C10) on either side of the power transformer to reduce EMI. Snubber circuitry reduces the high-frequency switching noise caused by the bridge power rectifier. A standard electrolytic capacitor bank filters the rectifier output. Broadband low-noise linear voltage regulators regulate the capacitor bank output to $\pm 5\,V_{DC}$. The two DMTD sections are powered via separate $\pm 5\,V_{DC}$ regulators for improved isolation.

C.1.6 Data Recording

The HP53131A universal TIC has a GPIB interface for control and data recording. However, no GBIP-to-PC interface converters were available. Hence, Bangert's [134] idea of reading the TIC data from its printer port was used instead. Moreover, the TIC 'prints' the measured data via a two-wire serial interface which could be read using a standard serial port. The UCT DMTD data recording software can simultaneously record and real-time plot data from two HP53131A TICs. See Figure C.2c for a screenshot. The time interval error is plotted on three different time scales namely a minute, an hour and a day. However, it would be beneficial to add data statistics and also real-time plots of the ODEV, MDEV, TDEV and MTIE in the future.

C.2 Measured Noise Floor

This section presents the measured noise floor data for the UCT DMTD system. Calibration is necessary to confirm the correct operation and the performance limits. First, the two HP53131A TICs are calibrated. Then, the combined noise floor of the UCT DMTD is measured using the HP53131A TICs and the Agilent E4400B LO.

C.2.1 HP53131A Noise Floor

The HP53131A universal TIC has a datasheet specified resolution of 0.5 ns. This amounts to a DMTD resolution of 0.5 fs for a heterodyne factor of $R = 10^6$. Both HP53131A's at UCT have the relatively poor 'standard option' internal timebase. Therefore, both were referenced to GPS using a Meinberg M400 GPSDO. The TICs are calibrated by feeding a DMTD ZCD generated 10 Hz pulse to the start channel of each TIC. In both TICs, the 'common' input option is selected connecting the start and stop channels internally. Both TICs were offset by 450 μs. Now, the measured time difference is the TIC noise after subtracting the 450 μs offset. The two TICs performed nearly identically and as expected. Figure C.3a compares the timing noise floor of the two TICs over a 95 hour period. Figure C.3b compares the frequency stability of the two TICs. The data statistics are summarised in Table C.2.

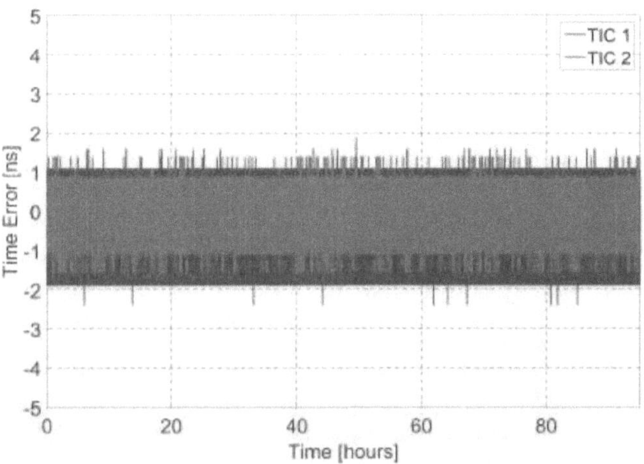

(a) Time Error versus Time Elapsed. TIC1 performed slightly worse than TIC2, with peak-to-peak timing errors of 4 ns and 3.5 ns,respectively. The mean time offsets for TIC1 and TIC2 were -0.558 ns and -0.225 ns, respectively.

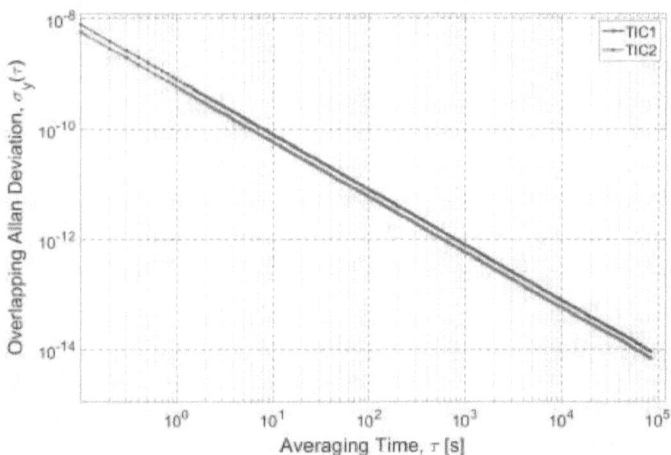

(b) Overlapping Allan Deviation. The TICs performed similarly with TIC2 performing slightly better. The ODEV of both has a constant slope of -1 confirmed to be pure white PM noise. After one second of averaging, TIC1 and TIC2 reached an ODEV of 7.96×10^{-10} and 5.84×10^{-10}, respectively.

Figure C.3: Measured HP53131A Universal TIC Noise Floor.

C.2.2   UCT DMTD Noise Floor

The noise floor of the DMTD (Section A[1]) is measured by feeding identical signals to both RF inputs. Thus, the measured time difference is the DMTD noise contribution. Moreover, a 10 MHz signal is split 1:2 using the front panel splitter and fed to both DMTD RF inputs. Flexible RG-178 cables were used but kept short ($\approx$ 30 cm) to limit temperature induced phase shifts. The Agilent E4400B LO generated an offset 10 MHz to produce the 10 Hz beat. The two TICs and LO were both GPS-referenced using a Meinberg M400 GPSDO.

No attempt was made to calibrate the absolute time offset within the DMTD system. However, after subtracting the TIC offsets the measured offset was about 200 ps with the cable lengths matched to within 1 cm. The cable propagation delay is about 33 ps/cm, and it is reasonable to approximate the internal DMTD time offset as 170 ps. The measured timing noise is plotted in Figure C.4a. The time drifted by approximately 30 ps during the 60-hour measurement. A somewhat poor quadratic polynomial fit suggests a negligible frequency offset of about -1.93×10^{-17} and a negligible daily ageing rate of -3.5×10^{-19}. Further investigation is required to find the cause of this drift. A better fit higher-order polynomial was subtracted to produce the residual time error in Figure C.4b. This residual had a peak-to-peak time error of 10.8 ps with a standard deviation of 0.73 ps over 60 hours. Figure C.5 plots the frequency noise. The frequency offset and drift are negligible, and the red lines indicate the ± 5 MAD boundaries at $\pm 6.19 \times 10^{-14}$. Some outliers are visible crossing the ± 5 MAD boundaries. The measured frequency stability is plotted in Figure C.7. As one would expect, the frequency stability improves roughly linearly with increasing averaging periods. The ODEV reaches 1.65×10^{-13} after one second of averaging and drops below 10^{-14} after approximately 200 seconds. An MDEV analysis identified mostly White PM and White FM noise. There are also components of Flicker PM noise. The time stability in Figure C.6 crosses 1 ps after about 14 hours of averaging. The data statistics are summarised in Table C.2. A full characterisation of the DMTD (Section B), as well as, the cross-channel and the cross-correlation noise floor is left as future work.

The goal is achieved with the DMTD noise floor is well below the expected UCT GPSDO noise floor. Moreover, the noise floor is adequate to characterise any current

---

[1]Section B performed similar to A but with a slightly higher noise level. A manufacturing issue in RF3 channel seems to be the cause of the extra noise. However, both channel A and B perform satisfactorily, and the repair is left for future work.

Table C.2: UCT DMTD Noise Floor Specifications

Description	Input	Beat	Offset	Pk-Pk	$\text{ODEV}_{\tau=1\text{s}}$
TIC1	10 Hz	-	$-0.558\,\text{ns}$	4 ns	7.96×10^{-10}
TIC2	10 Hz	-	$-0.225\,\text{ns}$	3.5 ns	5.84×10^{-10}
DMTD	10 MHz	10 Hz	$\approx 170\,\text{ps}$	10.8 ps	1.65×10^{-13}

or immediate future UCT frequency standard.

However, the UCT DMTD did perform somewhat worse than the Riley design [117] it is based on. In hindsight, the E4400B might have been a poor LO choice. Moreover, the UCT E4400B has the 'standard' option internal time base with a specified ageing rate of 1×10^{-6}/day. In comparison, the average performing Oscilloquartz 8788 OCXO has an ageing rate of 5×10^{-10}/day and 7×10^{-8}/year. This suggests that the 'standard' option internal timebase has poor frequency stability. A poor LO frequency stability along with a non-zero RF time difference limits the noise floor of the DMTD system [127].

A more stable LO and a variable delay line to null the RF phase offset is expected to lower the DMTD system noise floor. At high RF input levels, extra attenuation pads should improve the VSWR stability [126]. At higher RF frequencies, cable lengths should be kept equal to one half the wavelength for improved VSWR stability [126]. Additional future performance enhancements may include optimisation of the ZCD gain and bandwidth settings, lowering the ZCD gain by a factor of ten, using rigid internal RF cables, using a copper thermocouple, and a lab with less temperature variation.

C.3 Conclusion

This appendix detailed the UCT DMTD design and calibrated its noise floor. A low-cost four channel DMTD was designed capable of measuring 1 MHz to 100 MHz oscillators at beat frequencies of 1 Hz to 100 Hz. The system uses an external LO and TICs which makes it simple but more versatile. It can compare up to four oscillators simultaneously and is capable of ultra-low noise cross-correlation and three-cornered hat measurements. Two HP53131A universal TICs were used and an

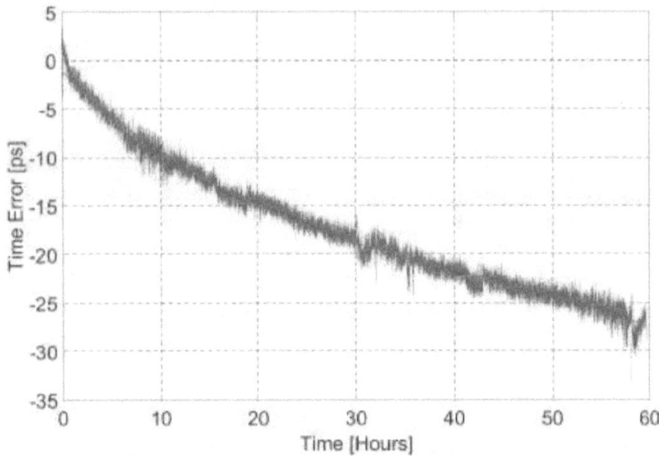

(a) Time Error versus Time Elapsed. The time drifted by approximately 30 ps during the 60-hour measurement. A quadratic polynomial fit suggests a frequency offset of about -1.93×10^{-17} and a daily ageing rate of -3.5×10^{-19}. Further investigation is required to find the cause of this drift.

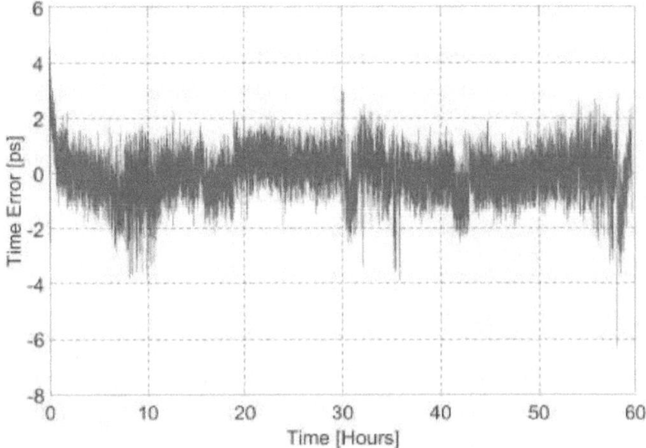

(b) Time Error versus Time Elapsed (polynomial fit removed). The residual peak-to-peak time error over 60 hours is 10.8 ps with a standard deviation of 0.73 ps.

Figure C.4: UCT DMTD Timing Noise Floor.

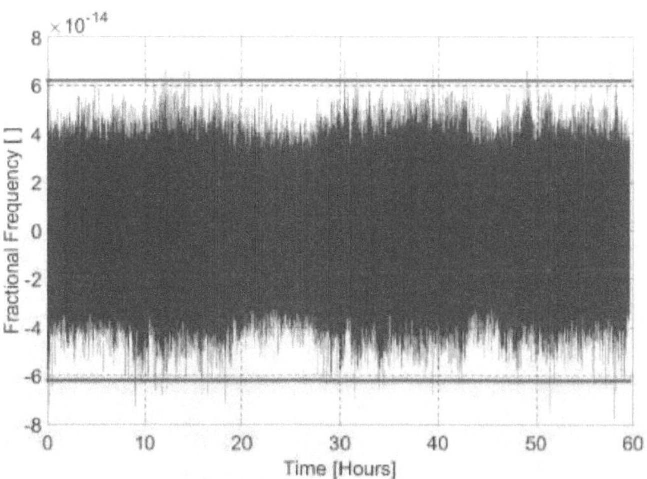

Figure C.5: UCT DMTD Frequency Noise Floor. The frequency offset and drift are negligible, and the red lines indicate the ± 5 MAD boundaries at $\pm 619 \times 10^{-14}$. Some outliers are visible crossing the ± 5 MAD boundaries.

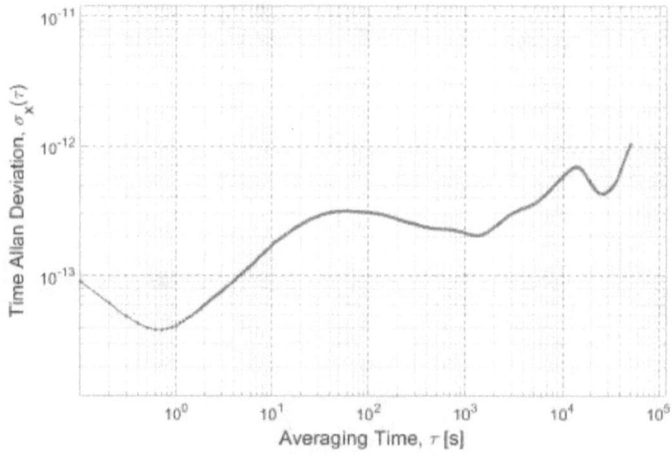

Figure C.6: UCT DMTD Time Stability. Time instability crosses 1 ps after about 14 hours of averaging.

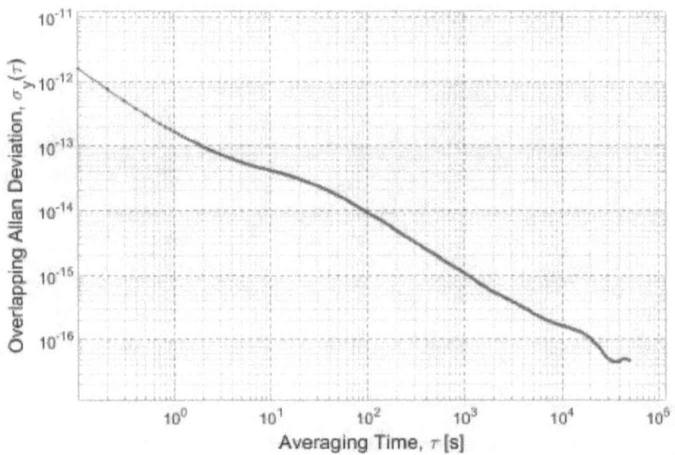

(a) Overlapping Allan Deviation. The frequency stability improves roughly linearly with increasing averaging periods. The ODEV reaches 1.65×10^{-13} after one second of averaging and drops below 10^{-14} after approximately 200 seconds.

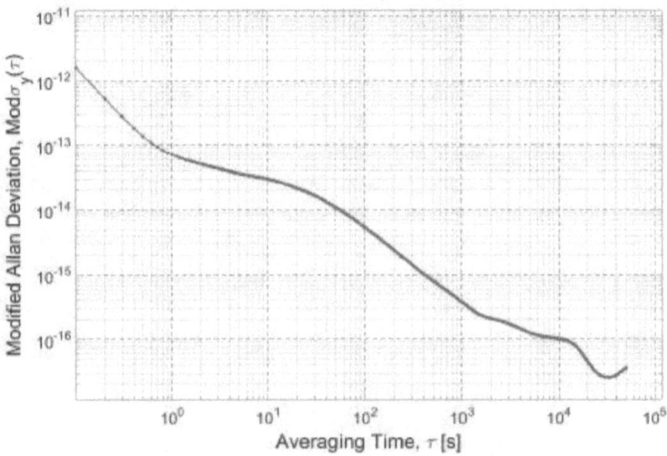

(b) Modified Allan Deviation. The noise types consist mostly of White PM and White FM. However, there are also components of Flicker PM noise.

Figure C.7: UCT DMTD Frequency Stability.

Agilent E4400B for the LO.

The two HP53131As were first calibrated. Then, the noise floor of the UCT DMTD was measured using the HP53131A TICs and the Agilent E4400B as LO. The system achieved femtosecond resolution and a 10 MHz ODEV noise floor of 1.65×10^{-13} at one second for a 10 Hz beat. The time drifted by 30 ps over a 60 hour period. This noise floor is well below our current and near future requirements. However, the noise floor may be lowered further by swapping the Agilent E4400B with a more stable LO.

Future upgrades to the DMTD may include improvements to the user interface, optimisation of the ZCDs, rigid internal RF cables, and a copper mixer thermocouple. Also, an external variable phase delay device to null the input RF phase difference is highly desirable [127]. It will also be beneficial to characterise Section B, the cross-channels, and the cross-correlation noise floor in the future.

Future high-fidelity coherent bistatic radar may require carriers and mixing stages with high frequency stability. Much of the literature gathered in this appendix is applicable to such radars. Moreover, accurate impedance matching and VSWR stabilisation are essential [132]. The VSWR may be stabilised in many ways including additional attenuation [132], high amplifier reverse isolation [126], high mixer isolation [126, 132], proper mixer impedance matching [132], and cables and PCB tracks that are multiples of half a wavelength long [126]. Amplifier harmonic distortion affects frequency stability [126, 128], but a double balanced mixer cancels the even harmonics [128]. Filters should be designed with a flat phase response and build with dielectrics that have small temperature coefficients. [129] Cables are sensitive to vibration and temperature and should be rigid and short [126, 129]. Mixers are phase sensitive and may require temperature stabilisation [132]. The use of voltage step-up transformers at amplifier inputs is desirable since the signal and noise power increase by the same ratio keeping the spectral density of phase unchanged [129].

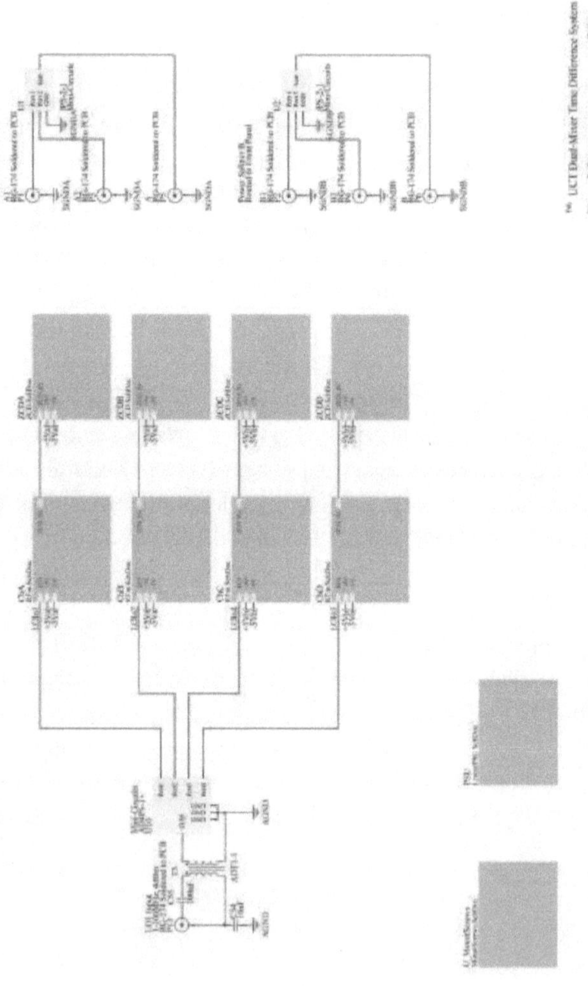

Figure C.8: DMTD Schematics 1 of 3

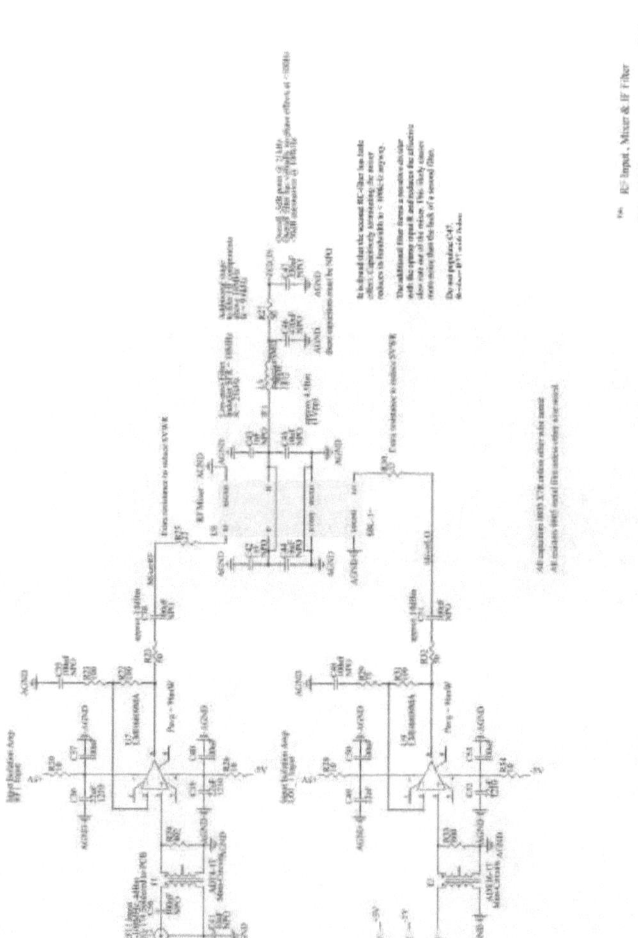

Figure C.9: DMTD Schematics 2 of 3

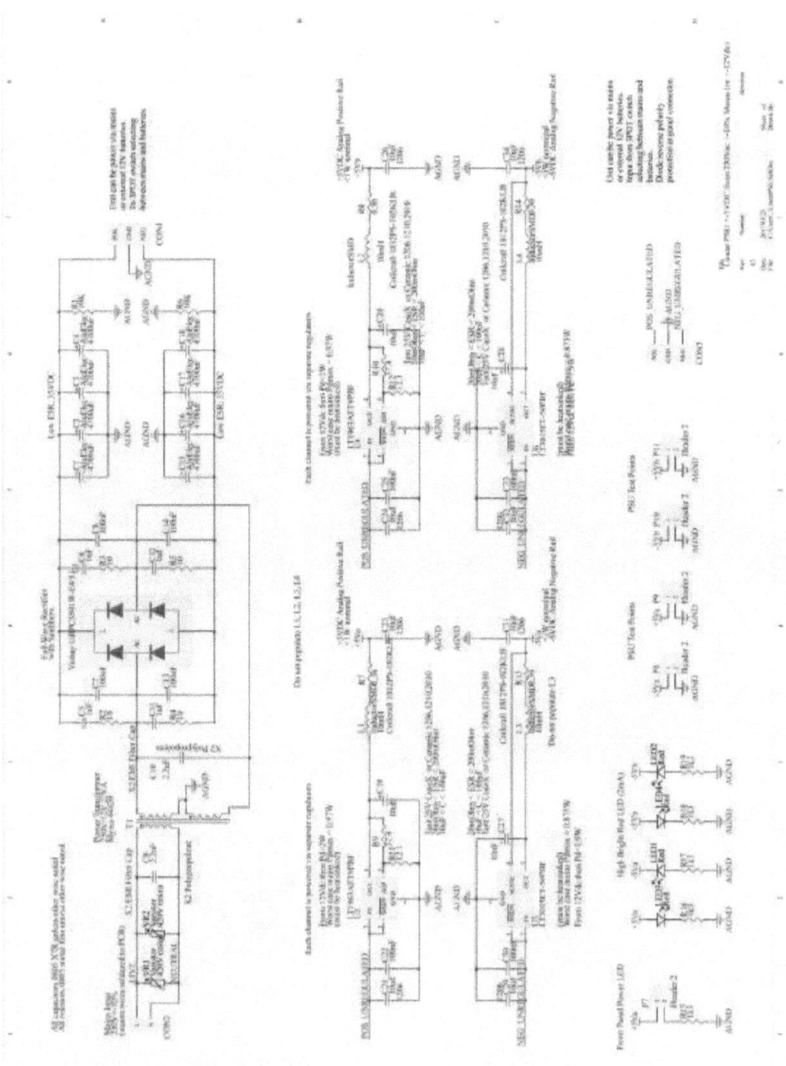

Figure C.10: DMTD Schematics 3 of 3

References

[1] N. J. Willis and H. Griffiths, Eds., *Advances in bistatic radar*, Raleigh, NC: SciTech Pub, 2007, 493 pp., ISBN: 978-1-891121-56-2.

[2] H. D. Griffiths, 'New directions in bistatic radar', in *2008 IEEE Radar Conference*, May 2008, pp. 1–6. DOI: 10.1109/RADAR.2008.4720719.

[3] N. Willis, *Bistatic radar*, ser. Artech House radar library. Artech House, 1991, ISBN: 9780890064276.

[4] C. J. Baker and A. L. Hume, 'Netted radar sensing', *IEEE Aerospace and Electronic Systems Magazine*, vol. 18, no. 2, pp. 3–6, Feb. 2003, ISSN: 0885-8985. DOI: 10.1109/MAES.2003.1183861.

[5] C. J. Baker, 'Cognitive diversity sensing', in *2010 International Conference on Electromagnetics in Advanced Applications*, Sep. 2010, pp. 875–878. DOI: 10.1109/ICEAA.2010.5653995.

[6] H. D. Griffiths, 'From a different perspective: Principles, practice and potential of bistatic radar', in *2003 Proceedings of the International Conference on Radar (IEEE Cat. No.03EX695)*, Sep. 2003, pp. 1–7. DOI: 10.1109/RADAR.2003.1278701.

[7] G. Retzer, 'A concept for signal processing in bistatic radar', presented at the IEEE International Radar Conference, Arlington, VA: IEEE, 1980, pp. 288–293. [Online]. Available: http://adsabs.harvard.edu/abs/1980inra.conf..288R (visited on 13/02/2018).

[8] J. C. Kirk Jr., 'Bistatic SAR motion compensation', presented at the IEEE International Radar Conference, IEEE, 1985, pp. 360–365. [Online]. Available: http://adsabs.harvard.edu/abs/1985inra.conf..360K (visited on 13/02/2018).

[9] C. K. Bovey and C. P. Horne, 'Synchronization aspects for bistatic radars', presented at the IEE International Radar Conference, IEE, 1987, pp. 22–25. [Online]. Available: `http://adsabs.harvard.edu/abs/1987rapr.conf...22B` (visited on 13/02/2018).

[10] J. L. Auterman, 'Phase stability requirements for a bistatic SAR system', presented at the IEEE International Radar Conference, Atlanta, GA: IEEE, 1984, pp. 48–52.

[11] G. Krieger, H. Fiedler, M. Zink, I. Hajnsek, M. Younis, S. Huber, M. Bachmann, J. H. Gonzalez, M. Werner and A. Moreira, 'TanDEM-X: A satellite formation for high-resolution SAR interferometry', in *2007 IET International Conference on Radar Systems*, Oct. 2007, pp. 1–5.

[12] T. E. Derham, 'Design and evaluation of a coherent multistatic radar system.', Doctoral, University of London, 2005, 329 pp. [Online]. Available: `http://discovery.ucl.ac.uk/1446245/` (visited on 31/12/2017).

[13] S. R. Doughty, 'Development and performance evaluation of a multistatic radar system', Doctoral, University of London, 2008, 213 pp. [Online]. Available: `http://discovery.ucl.ac.uk/14634/` (visited on 31/12/2017).

[14] L. Kesheng, 'An analysis of some problems of bistatic and multistatic radars', in *2003 Proceedings of the International Conference on Radar (IEEE Cat. No.03EX695)*, Sep. 2003, pp. 429–432. DOI: `10.1109/RADAR.2003.1278780`.

[15] M. R. B. Dunsmore, 'Bistatic radars for air defence', presented at the IEE International Radar Conference, IEE, 1987, pp. 7–11. [Online]. Available: `http://adsabs.harvard.edu/abs/1987rapr.conf....7D` (visited on 13/02/2018).

[16] T. A. Soame and D. M. Gould, 'Description of an experimental bistatic radar system', presented at the IEE International Radar Conference, IEE, 1987, pp. 12–16. [Online]. Available: `http://adsabs.harvard.edu/abs/1987rapr.conf...12S` (visited on 13/02/2018).

[17] J. Wurman, M. Randall, C. L. Frush, E. Loew and C. L. Holloway, 'Design of a bistatic dual-doppler radar for retrieving vector winds using one transmitter and a remote low-gain passive receiver', *Proceedings of the IEEE*, vol. 82, no. 12, pp. 1861–1872, Dec. 1994, ISSN: 0018-9219. DOI: `10.1109/5.338075`.

[18] M. A. Lombardi and A. N. Novick, 'Comparison of the One-Way and Common-View GPS Measurement Techniques Using a Known Frequency Offset', in *34th Annual Precise Time and Time Interval (PTTI) Meeting*, 2000, pp. 39–51.

[19] M. A. Lombardi, L. M. Nelson, A. N. Novick and V. S. Zhang, 'Time and frequency measurements using the global positioning system', *Cal Lab: International Journal of Metrology*, vol. 8, no. 3, pp. 26–33, 2001.

[20] M. Weiss, 'Synchronisation of bistatic radar systems', in *IGARSS 2004. 2004 IEEE International Geoscience and Remote Sensing Symposium*, vol. 3, Sep. 2004, 1750–1753 vol.3. DOI: 10.1109/IGARSS.2004.1370671.

[21] R. Saini, R. Zuo and M. Cherniakov, 'Problem of signal synchronisation in space-surface bistatic synthetic aperture radar based on global navigation satellite emissions - experimental results', *Sonar Navigation IET Radar*, vol. 4, no. 1, pp. 110–125, Feb. 2010, ISSN: 1751-8784. DOI: 10.1049/iet-rsn.2008.0121.

[22] M. A. Lombardi, 'The use of GPS disciplined oscillators as primary frequency standards for calibration and metrology laboratories', *NCSLI Measure*, vol. 3, no. 3, pp. 56–65, Sep. 2008, ISSN: 1931-5775, 2381-0580. DOI: 10.1080/19315775.2008.11721437. [Online]. Available: https://www.tandfonline.com/doi/full/10.1080/19315775.2008.11721437 (visited on 07/08/2018).

[23] T. E. Parker and D. Matsakis, 'Time and Frequency Dissemination: Advances in GPS Time Transfer Techniques', *GPS World*, Nov. 2004.

[24] D. Allan and J. Levine, 'A rubidium frequency standard and a GPS receiver. a remotely steered clock system with good short-term and long-term stability', IEEE, 1990, pp. 151–160. DOI: 10.1109/FREQ.1990.177493. [Online]. Available: http://ieeexplore.ieee.org/document/177493/ (visited on 05/08/2018).

[25] K. Weiss, A. Lisowiec and W. Steplewski, 'Limit parameters of GPS disciplined quartz frequency sources', in *First International Conference on Advanced Optoelectronics and Lasers, 2003. Proceedings of CAOL 2003*, vol. 2, Sep. 2003, 262–264 vol.2. DOI: 10.1109/CAOL.2003.1251326.

[26] U. Government, Ed. (2000). GPS.GOV: Selective availability, [Online]. Available: https://www.gps.gov/systems/gps/modernization/sa/ (visited on 17/10/2017).

[27] M. Inggs, H. Griffiths, K. Woodbridge, W. Al-Ashwal, J. Sandenbergh and G. Inggs, 'Experimental networked radar using GPS-disciplined reference oscillators', in *IET COGnitive systems with Interactive Sensors (GOGIS)*, 2010.

[28] W. Bing, Y. Wei, Y. Dongkai and Z. Qishan, 'A new method of phase synchronization for GPS signals based bistatic radar', in *World Automation Congress 2012*, Jun. 2012, pp. 1–4.

[29] T. E. Derham, S. Doughty, K. Woodbridge and C. J. Baker, 'Design and evaluation of a low-cost multistatic netted radar system', *Sonar Navigation IET Radar*, vol. 1, no. 5, pp. 362–368, Oct. 2007, ISSN: 1751-8784. DOI: 10.1049/iet-rsn:20060100.

[30] C. Baojian, Z. Dehai and X. Dazhi, 'An improved time synchronous system based on GPS disciplined rubidium', in *2010 International Conference on Intelligent Computation Technology and Automation*, vol. 1, May 2010, pp. 599–602. DOI: 10.1109/ICICTA.2010.843.

[31] M. A. Lombardi, A. N. Novick and V. S. Zhang, 'Characterizing the performance of GPS disciplined oscillators with respect to UTC(NIT)', in *Proceedings of the 2005 IEEE International Frequency Control Symposium and Exposition, 2005.*, Aug. 2005, pp. 677–684. DOI: 10.1109/FREQ.2005.1574017.

[32] S. J. Goldman, *Phase noise analysis in radar systems using personal computers.* New York: Wiley, 1989, 518 pp., ISBN: 978-0-471-61894-2.

[33] S. Alhuwaimel, S. Coetzee, P. Cheng, D. D. Plessis, F. Fioranelli, H. Griffiths, M. R. Inggs, D. Jordan, W. Miceli, D. O'Hagan, R. Palama, S. Paine, M. Ritchie, J. S. Sandenbergh and A. Stevens, 'First measurements with NeXtRAD, a polarimetric X/L band radar network', in *2017 IEEE Radar Conference (RadarConf)*, May 2017, pp. 1663–1668. DOI: 10.1109/RADAR.2017.7944474.

[34] J. S. Sandenbergh and M. R. Inggs, 'A common view GPSDO to synchronize netted radar', in *2007 IET International Conference on Radar Systems*, Oct. 2007, pp. 1–5. DOI: 10.1049/cp:20070499.

[35] J. Sandenbergh and M. Inggs, 'Synchronizing network radar using all-in-view GPS-disciplined oscillators', in *2017 IEEE Radar Conference (RadarConf)*, May 2017, pp. 1640–1645. DOI: 10.1109/RADAR.2017.7944470.

[36] S. M. Hurley, M. Tummala, T. O. Walker and P. E. Pace, 'Impact of synchronization on signal-to-noise ratio in a distributed radar system', in *2006 IEEE/SMC International Conference on System of Systems Engineering*, Apr. 2006, pp. 24–26. DOI: 10.1109/SYSOSE.2006.1652311.

[37] W. Bing-Yang and M. Yan-Hcng, 'Analysis of detection accuracy of bistatic radar system', in *2016 2nd IEEE International Conference on Computer and Communications (ICCC)*, Oct. 2016, pp. 1828–1831. DOI: 10.1109/CompComm.2016.7925018.

[38] G. Krieger and M. Younis, 'Impact of oscillator noise in bistatic and multistatic SAR', *IEEE Geoscience and Remote Sensing Letters*, vol. 3, no. 3, pp. 424–428, Jul. 2006, ISSN: 1545-598X. DOI: 10.1109/LGRS.2006.874164.

[39] J. F. Cliche and B. Shillue, 'Precision timing control for radioastronomy: Maintaining femtosecond synchronization in the atacama large millimeter array', *IEEE Control Systems*, vol. 26, no. 1, pp. 19–26, Feb. 2006, ISSN: 1066-033X. DOI: 10.1109/MCS.2006.1580149.

[40] J.-D. Deschenes, L. C. Sinclair, F. R. Giorgetta, W. C. Swann, E. Baumann, H. Bergeron, M. Cermak, I. Coddington and N. R. Newbury, 'Synchronization of distant optical clocks at the femtosecond level', *Physical Review X*, vol. 6, no. 2, 11th May 2016, ISSN: 2160-3308. DOI: 10.1103/PhysRevX.6.021016. [Online]. Available: https://link.aps.org/doi/10.1103/PhysRevX.6.021016 (visited on 10/04/2018).

[41] S. Lewis, J. Sandenbergh and M. Inggs, 'Evaluating an off-the-shelf white rabbit system to synchronise network radar via optic fibre', in *2017 IEEE Radar Conference (RadarConf)*, May 2017, pp. 1657–1662. DOI: 10.1109/RADAR.2017.7944473.

[42] M. R. Inggs, J. S. Sandenbergh and S. A. C. Lewis, 'Investigation of white rabbit for synchronization and timing of netted radar', in *2015 IEEE Radar Conference*, Oct. 2015, pp. 214–217. DOI: 10.1109/RadarConf.2015.7411882.

[43] N. Kaur, F. Frank, P. E. Pottie and P. Tuckey, 'Time and frequency transfer over a 500 km cascaded white rabbit network', in *2017 Joint Conference of the European Frequency and Time Forum and IEEE International Frequency Control Symposium (EFTF/IFCS)*, Jul. 2017, pp. 86–90. DOI: 10.1109/FCS.2017.8088808.

[44] E. F. Dierikx, A. E. Wallin, T. Fordell, J. Myyry, P. Koponen, M. Merimaa, T. J. Pinkert, J. C. J. Koelemeij, H. Z. Peek and R. Smets, 'White rabbit precision time protocol on long-distance fiber links', *IEEE Transactions on Ultrasonics, Ferroelectrics, and Frequency Control*, vol. 63, no. 7, pp. 945–952, Jul. 2016, ISSN: 0885-3010. DOI: 10.1109/TUFFC.2016.2518122.

[45] D. Kirchner, 'Two-way time transfer via communication satellites', *Proceedings of the IEEE*, vol. 79, no. 7, pp. 983–990, Jul. 1991, ISSN: 0018-9219. DOI: 10.1109/5.84975.

[46] R. L. Schmid, T. M. Comberiate, J. E. Hodkin and J. A. Nanzer, 'A distributed RF transmitter using one-way wireless clock transfer', *IEEE Microwave and Wireless Components Letters*, vol. 27, no. 2, pp. 195–197, Feb. 2017, ISSN: 1531-1309, 1558-1764. DOI: 10.1109/LMWC.2017.2648510. [Online]. Available: http://ieeexplore.ieee.org/document/7839937/ (visited on 23/03/2018).

[47] M. Younis, R. Metzig and G. Krieger, 'Performance prediction of a phase synchronization link for bistatic SAR', *IEEE Geoscience and Remote Sensing Letters*, vol. 3, no. 3, pp. 429–433, Jul. 2006, ISSN: 1545-598X. DOI: 10.1109/LGRS.2006.874163.

[48] W.-Q. Wang and H. Shao, 'Performance prediction of a synchronization link for distributed aerospace wireless systems', *The Scientific World Journal*, vol. 2013, 22nd Jul. 2013, ISSN: 2356-6140. DOI: 10.1155/2013/159742. [Online]. Available: https://www.ncbi.nlm.nih.gov/pmc/articles/PMC3736480/ (visited on 13/02/2018).

[49] W. Q. Wang, 'Carrier frequency synchronization in distributed wireless sensor networks', *IEEE Systems Journal*, vol. 9, no. 3, pp. 703–713, Sep. 2015, ISSN: 1932-8184. DOI: 10.1109/JSYST.2014.2330392.

[50] M. Rico, J. P. Aubry, C. Botteron and P. A. Farine, 'Ns-Level time transfer over a microwave link using the PTP-WR protocol', in *2015 Joint Conference of the IEEE International Frequency Control Symposium the European Frequency and Time Forum*, Apr. 2015, pp. 690–695. DOI: 10.1109/FCS. 2015.7138936.

[51] Z. He, G. Jin, H. Feng, Q. Zhang, Z. Dong, H. Huang and G. Chen, 'Phase synchronization method for distributed spaceborne fmcw SAR system', in *2016 IEEE International Geoscience and Remote Sensing Symposium (IG-ARSS)*, Jul. 2016, pp. 7376–7379. DOI: 10.1109/IGARSS.2016.7730924.

[52] D. D. Thomas, 'Synchronization of noncooperative bistatic radar receivers', Doctoral, Syracuse University, New York, 1999. [Online]. Available: https://surface.syr.edu/eecs_etd/165/ (visited on 13/02/2018).

[53] C. S. Pappu, B. C. Flores, P. S. Debroux and J. E. Boehm, 'An electronic implementation of lorenz chaotic oscillator synchronization for bistatic radar applications', *IEEE Transactions on Aerospace and Electronic Systems*, vol. 53, no. 4, pp. 2001–2013, Aug. 2017, ISSN: 0018-9251. DOI: 10.1109/TAES.2017. 2680661.

[54] C. S. Pappu, B. C. Flores, P. S. Debroux, B. Verdin and J. Boehm, 'Synchronisation of bistatic radar using chaotic AM and chaos-based FM waveforms', *Sonar Navigation IET Radar*, vol. 11, no. 1, pp. 90–97, 2017, ISSN: 1751-8784. DOI: 10.1049/iet-rsn.2016.0043.

[55] E. Kaplan and C. Hegarty, *Understanding GPS: Principles and Applications*. Artech House, 2005, ISBN: 9781580538954.

[56] U. Government, Ed. (2004). GPS.GOV: U.S. space-based positioning, navigation, and timing (PNT) policy of 2004, [Online]. Available: https://www. gps.gov/policy/docs/2004/ (visited on 17/10/2017).

[57] G. Petit and C. Thomas, 'GPS frequency transfer using carrier phase measurements', in *Proceedings of 1996 IEEE International Frequency Control Symposium*, Jun. 1996, pp. 1151–1158. DOI: 10.1109/FREQ.1996.560307.

[58] U. Government, Ed. (2007). Statement by the press secretary, [Online]. Available: https://georgewbush-whitehouse.archives.gov/news/releases/ 2007/09/20070918-2.html (visited on 17/10/2017).

[59] 'M12+ GPS Receiver User's Guide', Motorola (Synergy Systems), User Manual, 2004. [Online]. Available: `https://www.cnssys.com/files/M12+UsersGuide.pdf` (visited on 16/10/2017).

[60] R. M. Hambly and T. A. Clark, 'Critical evaluation of the Motorola M12+ GPS timing receiver vs. the master clock at the United States Naval Observatory, Washington, DC', in *34th Annual Precise Time and Time Interval (PPTI) Meeting*, VA, 2002, pp. 109–116.

[61] A. Lisowiec and L. Nafalski, 'Progress in temperature stabilization of GPS antennas', in *2004 18th European Frequency and Time Forum (EFTF 2004)*, Apr. 2004, pp. 134–136. DOI: `10.1049/cp:20040835`.

[62] L. Gun, W. Fu-ping, W. Jing-fa and H. Xian-he, 'Development of high precision and multifunctional timing system using integrated GPS/BD receiver', in *2008 IEEE International Frequency Control Symposium*, May 2008, pp. 507–514. DOI: `10.1109/FREQ.2008.4623051`.

[63] D. W. Allan, L. Fey, H. E. Machlan and J. A. Barnes, 'An UltraPrecise time synchronization system designed by computer simulation', *Frequency*, pp. 11–14, 1968.

[64] R. Hardin and M. Yankowski, 'Performance of a PID phase lock loop with kalman filtered input data [GPS data]', in *Proceedings of the 1992 IEEE Frequency Control Symposium*, May 1992, pp. 238–256. DOI: `10.1109/FREQ.1992.270008`.

[65] C. Little and C. Green, 'GPS disciplined rubidium oscillator', in *Tenth European Frequency and Time Forum EFTF 96 (IEE Conf. Publ. 418)*, Mar. 1996, pp. 105–110. DOI: `10.1049/cp:19960026`.

[66] A. Lisowiec, A. Czarnecki and Z. Rau, 'A method for quartz oscillator synchronization by GPS signal', in *Tenth European Frequency and Time Forum EFTF 96 (IEE Conf. Publ. 418)*, Mar. 1996, pp. 96–99. DOI: `10.1049/cp:19960024`.

[67] B. M. Penrod, 'Adaptive temperature compensation of GPS disciplined quartz and rubidium oscillators', in *Proceedings of 1996 IEEE International Frequency Control Symposium*, Jun. 1996, pp. 980–987. DOI: `10.1109/FREQ.1996.560284`.

[68] P. Rochat, B. Leuenberger and X. Stehlin, 'A new synchronized ultra miniature rubidium oscillator', in *Proceedings of the 2002 IEEE International Frequency Control Symposium and PDA Exhibition (Cat. No.02CH37234)*, 2002, pp. 451–454. DOI: 10.1109/FREQ.2002.1075924.

[69] C.-L. Cheng, F.-R. Chang and K.-Y. Tu, 'Highly accurate real-time GPS carrier phase-disciplined oscillator', *IEEE Transactions on Instrumentation and Measurement*, vol. 54, no. 2, pp. 819–824, Apr. 2005, ISSN: 0018-9456. DOI: 10.1109/TIM.2004.843403.

[70] H. Yulin, Y. Jianyu and X. Jintao, 'Synchronization technology of bistatic radar system', in *2006 International Conference on Communications, Circuits and Systems*, vol. 4, Jun. 2006, pp. 2219–2221. DOI: 10.1109/ICCCAS.2006.285118.

[71] B. Cui, X. Hou and D. Zhou, 'Methodological approach to GPS disciplined OCXO based on PID PLL', in *2009 9th International Conference on Electronic Measurement Instruments*, Aug. 2009, pp. 1-528-1-533. DOI: 10.1109/ICEMI.2009.5274810.

[72] T. McClelland, I. Shtaerman, E. Zarjetski, R. Baransky and M. Khurgin, 'Disciplined rubidium oscillator for harsh environments', in *2011 Joint Conference of the IEEE International Frequency Control and the European Frequency and Time Forum (FCS) Proceedings*, May 2011, pp. 1–5. DOI: 10.1109/FCS.2011.5977905.

[73] C. Green, 'Advantageous GPS disciplining of the OSA BVA oscillator', in *2012 European Frequency and Time Forum*, Apr. 2012, pp. 263–266. DOI: 10.1109/EFTF.2012.6502379.

[74] B. Shera, 'A GPS-based frequency standard', pp. 37–444, Jul. 1998.

[75] T. Van Baak. (May 2007). Jackson labs fury GPS disciplined oscillator, Leapsecond.com, [Online]. Available: http://www.leapsecond.com/pages/fury/ (visited on 10/09/2018).

[76] W. Q. Wang, 'GPS-based time phase synchronization processing for distributed SAR', *IEEE Transactions on Aerospace and Electronic Systems*, vol. 45, no. 3, pp. 1040–1051, Jul. 2009, ISSN: 0018-9251. DOI: 10.1109/TAES.2009.5259181.

[77] W. F. Egan, *Frequency synthesis by phase lock*, 2nd ed. New York: Wiley, 2000, 597 pp., ISBN: 978-0-471-32104-0.

[78] ——, *Phase-lock basics*, 2nd ed. Hoboken, N.J. : New York: Wiley-Interscience ; IEEE, 2008, 441 pp., OCLC: ocn122338167, ISBN: 978-0-470-11800-9.

[79] D. H. Wolaver, *Phase-locked loop circuit design*, ser. Prentice Hall biophysics and bioengineering series. Englewood Cliffs, N.J: Prentice Hall, 1991, 262 pp., ISBN: 978-0-13-662743-2.

[80] J. Vila-Valls, P. Closas, M. Navarro and C. Fernandez-Prades, 'Are PLLs dead? a tutorial on kalman filter-based techniques for digital carrier synchronization', *IEEE Aerospace and Electronic Systems Magazine*, vol. 32, no. 7, pp. 28–45, Jul. 2017, ISSN: 0885-8985. DOI: 10.1109/MAES.2017.150260.

[81] F. Cordara and V. Pettiti, 'Short-term characterization of GPS disciplined oscillators and field trial for frequency of italian calibration centres', in *Proceedings of the 1999 Joint Meeting of the European Frequency and Time Forum and the IEEE International Frequency Control Symposium*, vol. 1, 1999, 404–407 vol.1. DOI: 10.1109/FREQ.1999.840792.

[82] Y. Kim and C. P. Walter, 'Retrace and disciplining time constant effects on holdover clock drifts in chip-scale atomic clock', in *2017 Joint Conference of the European Frequency and Time Forum and IEEE International Frequency Control Symposium (EFTF/IFCS)*, Jul. 2017, pp. 310–314. DOI: 10.1109/FCS.2017.8088878.

[83] P. Eskelinen, 'Observations on stability measurements of commercial atomic clocks', in *Proceedings of the 1999 Joint Meeting of the European Frequency and Time Forum and the IEEE International Frequency Control Symposium (Cat. No.99CH36313)*, vol. 1, 1999, 186–189 vol.1. DOI: 10.1109/FREQ.1999.840739.

[84] T. Johnsen, 'Time and frequency synchronization in multistatic radar. consequences to usage of GPS disciplined references with and without GPS signals', in *Proceedings of the 2002 IEEE Radar Conference (IEEE Cat. No.02CH37322)*, 2002, pp. 141–147. DOI: 10.1109/NRC.2002.999711.

[85] J. D. Sahr and F. D. Lind, 'The manastash ridge radar: A passive bistatic radar for upper atmospheric radio science', *Radio Science*, vol. 32, no. 6, pp. 2345–2358, Nov. 1997, ISSN: 1944-799X. DOI: 10.1029/97RS02454.

[86]   Y. Wei, P. Tong, R. Xu and L. Yu, 'Experimental analysis of a HF hybrid sky-surface wave radar', *IEEE Aerospace and Electronic Systems Magazine*, vol. 33, no. 3, pp. 32–40, Mar. 2018, ISSN: 0885-8985. DOI: 10.1109/MAES.2018.170036.

[87]   D. W. Allan, 'Time and frequency (time-domain) characterization, estimation, and prediction of precision clocks and oscillators', *IEEE Transactions on Ultrasonics, Ferroelectrics, and Frequency Control*, vol. 34, no. 6, pp. 647–654, Nov. 1987, ISSN: 0885-3010. DOI: 10.1109/T-UFFC.1987.26997.

[88]   F. L. Walls, A. J. D. Clements, C. M. Felton, M. A. Lombardi and M. D. Vanek, 'Extending the range and accuracy of phase noise measurements', in *Proceedings of the 42nd Annual Frequency Control Symposium, 1988.*, Jun. 1988, pp. 432–441. DOI: 10.1109/FREQ.1988.27636.

[89]   S. Kodituwakku and H. T. Tran, *Detection of fast moving and accelerating targets compensating range and doppler migration*, Jun. 2014. [Online]. Available: http://www.dtic.mil/get-tr-doc/pdf?AD=ADA615319 (visited on 05/07/2018).

[90]   M. A. Richards, *Fundamentals of Radar Signal Processing, Second Edition*, 2 edition. New York: McGraw-Hill Education, 14th Jan. 2014, 656 pp., ISBN: 978-0-07-179832-7.

[91]   A. SenGupta and F. L. Walls, 'Effect of aliasing on spurs and PM noise in frequency dividers', in *Proceedings of the 2000 IEEE/EIA International Frequency Control Symposium and Exhibition (Cat. No.00CH37052)*, 2000, pp. 541–548. DOI: 10.1109/FREQ.2000.887414.

[92]   B. Griffiths. (2018). Clock shapers, Precision Time and Frequency Systems, [Online]. Available: http://www.ko4bb.com/~bruce/CLKSHPR.html (visited on 29/07/2018).

[93]   ——, (2018). Phase noise theory: Ideal frequency multipliers and dividers, Precision Time and Frequency Systems, [Online]. Available: http://www.ko4bb.com/~bruce/IdealFreqMultDiv.html (visited on 29/07/2018).

[94]   P. Tremblay and M. Tetu, 'Characterization of frequency stability: Effect of RF filtering', *IEEE Transactions on Instrumentation and Measurement*, vol. IM-34, no. 2, pp. 151–154, Jun. 1985, ISSN: 0018-9456. DOI: 10.1109/TIM.1985.4315290.

[95] 'PicoBlaze 8-Bit Embedded Microcontroller User Guide', Xilinx, User Manual UG129, 2011, p. 122. [Online]. Available: https : / / www . xilinx . com / support/documentation/ip_documentation/ug129.pdf (visited on 16/10/2017).

[96] J. G. Proakis and D. G. Manolakis, *Digital Signal Processing: Principles, Algorithms, and Applications*, 3rd. Prentice Hall, 1996, 968 pp., ISBN: 0-13-373762-4.

[97] B. P. Lathi and R. Green, '7.6 Frequency response of discrete-time systems', in *Essentials of digital signal processing*, Cambridge, UK: Cambridge University Press, 2014, p. 457, ISBN: 978-1-107-05932-0.

[98] D. C. LeBlanc, *Statistics: concepts and applications for science*. Boston: Jones and Bartlett, 2004, 382 pp., ISBN: 978-0-7637-4699-5.

[99] W. Riley, *Stable32 User Manual*. Hamilton Technical Services, 2008, 348 pp. [Online]. Available: http://www.stable32.com/Manual154.pdf (visited on 25/06/2018).

[100] F. Fioranelli, M. Ritchie, H. Griffiths, S. Sandenbergh and M. Inggs, 'Analysis of polarimetric bistatic sea clutter using the NetRAD radar system', *Sonar Navigation IET Radar*, vol. 10, no. 8, pp. 1356–1366, 2016, ISSN: 1751-8784. DOI: 10.1049/iet-rsn.2015.0416.

[101] M. Inggs, A. Balleri, W. A. Al-Ashwal, K. D. Ward, K. Woodbridge, M. Ritchie, W. Miceli, R. J. A. Tough, C. J. Baker, S. Watts, R. Harmanny, A. Stove, J. S. Sandenbergh and H. D. Griffiths, 'NetRAD multistatic sea clutter database', in *2012 IEEE International Geoscience and Remote Sensing Symposium*, Jul. 2012, pp. 2937–2940. DOI: 10.1109/IGARSS.2012.6350710.

[102] R. Palama, M. Ritchie, H. Griffiths, W. Miceli, F. Fioranelli, S. Sandenbergh and M. Inggs, 'Correlation analysis of simultaneously collected bistatic and monostatic sea clutter', in *2017 IEEE Radar Conference (RadarConf)*, May 2017, pp. 1466–1471. DOI: 10.1109/RADAR.2017.7944438.

[103] W. A. M. Al-Ashwal, 'Measurement and modelling of bistatic sea clutter', Doctoral, UCL (University College London), 28th Nov. 2011, 244 pp. [Online]. Available: http : / / discovery . ucl . ac . uk / 1334081/ (visited on 31/12/2017).

[104] W. A. Al-Ashwal, C. J. Baker, A. Balleri, H. D. Griffiths, R. Harmanny, M. Inggs, W. J. Miceli, M. Ritchie, J. S. Sandenbergh, A. Stove, R. J. A. Tough, K. D. Ward, S. Watts and K. Woodbridge, 'Statistical analysis of simultaneous monostatic and bistatic sea clutter at low grazing angles', *Electronics Letters*, vol. 47, no. 10, pp. 621–622, May 2011, ISSN: 0013-5194. DOI: 10.1049/el.2011.0557.

[105] M. Inggs, G. Inggs, S. Sandenbergh, W. Al-Ashwal, K. Woodbridge and H. Griffiths, 'Multistatic networked radar for sea clutter measurements', in *2011 IEEE International Geoscience and Remote Sensing Symposium*, Jul. 2011, pp. 4449–4452. DOI: 10.1109/IGARSS.2011.6050220.

[106] V. Carotenuto, A. Aubry, A. D. Maio and A. Farina, 'Phase noise modeling and its effects on the performance of some radar signal processors', in *2015 IEEE Radar Conference (RadarCon)*, May 2015, pp. 0274–0279. DOI: 10.1109/RADAR.2015.7131009.

[107] A. Aubry, V. Carotenuto, A. D. Maio and A. Farina, 'Radar phase noise modeling and effects-part II: Pulse doppler processors and sidelobe blankers', *IEEE Transactions on Aerospace and Electronic Systems*, vol. 52, no. 2, pp. 712–725, Apr. 2016, ISSN: 0018-9251. DOI: 10.1109/TAES.2015.140723.

[108] W. A. Al-Ashwal, A. Balleri, H. D. Griffiths, W. J. Miceli, K. Woodbridge, R. Harmanny, M. A. Ritchie, A. G. Stove, S. Watts, C. J. Baker, M. Inggs, J. S. Sandenbergh, R. J. A. Tough and K. D. Ward, 'Measurements of bistatic radar sea clutter', in *2011 IEEE RadarCon (RADAR)*, May 2011, pp. 217–221. DOI: 10.1109/RADAR.2011.5960531.

[109] J. Sandenbergh and M. Inggs, 'A summary of the results achieved by the GPS disciplined references of the NetRAD and NeXtRAD multistatic radars', in press, 2018.

[110] D. A. Howe, D. U. Allan and J. A. Barnes, 'Properties of signal sources and measurement methods', in *Thirty Fifth Annual Frequency Control Symposium*, May 1981, pp. 669–716. DOI: 10.1109/FREQ.1981.200541.

[111] D. Sullivan, D. Allan, D. Howe and F. Walls, Eds., *NIST Technical Note 1337: Characterization of Clocks and Oscillators*, NIST/TN-1337, Boulder, Colorado: National Institute of Standards and Technology, Mar. 1990, 352 pp.

[Online]. Available: `http://tf.boulder.nist.gov/general/pdf/868.pdf` (visited on 27/11/2017).

[112] S. Stein, 'Frequency and time - their measurement and characterization', in *Precision Frequency Control*, vol. 2, New York: Academic Press, 1985, pp. 191–232, 399–416.

[113] D. Allan, H. Hellwig, P. Kartaschoff, J. Vanier, J. Vig, G. M. R. Winkler and N. F. Yannoni, 'Standard terminology for fundamental frequency and time metrology', in *Proceedings of the 42nd Annual Frequency Control Symposium, 1988.*, Jun. 1988, pp. 419–425. DOI: `10.1109/FREQ.1988.27634`.

[114] W. Riley, *NIST Special Publication 1065: Handbook of Frequency*. Boulder, Colorado: National Institute of Standards and Technology, Jul. 2008, 136 pp. [Online]. Available: `http://nvlpubs.nist.gov/nistpubs/Legacy/SP/nistspecialpublication1065.pdf` (visited on 27/11/2017).

[115] J. Rutman, 'Characterization of phase and frequency instabilities in precision frequency sources: Fifteen years of progress', *Proceedings of the IEEE*, vol. 66, no. 9, pp. 1048–1075, Sep. 1978, ISSN: 0018-9219. DOI: `10.1109/PROC.1978.11080`.

[116] P. Bogdanov, A. Bandura, M. German, K. Kol'chenko and O. Nechaeva, 'Dual-frequency time transfer unit for comparisons of the remote clocks using GLONASS and GPS signals', in *2013 Joint European Frequency and Time Forum International Frequency Control Symposium (EFTF/IFC)*, Jul. 2013, pp. 505–507. DOI: `10.1109/EFTF-IFC.2013.6702144`.

[117] W. Riley, 'A small dual mixer time difference (DMTD) clock measuring system', Hamilton Technical Services, Technical Report, 2010. [Online]. Available: `http://www.stable32.com/A%20Small%20DMTD%20System.pdf` (visited on 16/10/2017).

[118] C. Greenhall, A. Kirk and R. Tjoelker, 'Frequency standards stability analyzer for the deep space network', JPL, Technical Report, 2007.

[119] L. Sojdr, J. Cermak and R. Barillet, 'Optimization of dual-mixer time-difference multiplier', pp. 588–594, 1st Jan. 2004. DOI: `10.1049/cp:20040934`. [Online]. Available: `http://digital-library.theiet.org/content/conferences/10.1049/cp_20040934` (visited on 17/10/2017).

[120] F. Nakagawa, M. Imae, Y. Hanado and M. Aida, 'Development of multi channel dual mixier time difference system', in *2004 Conference on Precision Electromagnetic Measurements*, Jun. 2004, pp. 234–235. DOI: 10.1109/CPEM.2004.305548.

[121] S. Stein, D. Glaze, J. Levine, J. Gray, D. Hilliard, D. Howe and L. A. Erb, 'Automated high-accuracy phase measurement system', *IEEE Transactions on Instrumentation and Measurement*, vol. 32, no. 1, pp. 227–231, Mar. 1983, ISSN: 0018-9456. DOI: 10.1109/TIM.1983.4315047.

[122] D. W. Allan, 'Report on NBS dual mixer time difference system (DMTD) built for time domain measurements associated with phase 1 of GPS', *NBS IR 75 827*, 1976. [Online]. Available: http://ci.nii.ac.jp/naid/10019514771/en/.

[123] D. W. Allan and H. Daams, 'Picosecond time difference measurement system', in *29th Annual Symposium on Frequency Control*, May 1975, pp. 404–411. DOI: 10.1109/FREQ.1975.200112.

[124] S. Stein, D. Glaze, J. Levine, J. Gray and D. Hilliard, 'Performance of an automated high accuracy phase measurement system,' National Bureau of Standards, Boulder, CO, Time and Frequency Division, 1982. [Online]. Available: http://www.dtic.mil/docs/citations/ADP001539 (visited on 17/10/2017).

[125] G. Brida, 'High resolution frequency stability measurement system', *Review of Scientific Instruments*, vol. 73, no. 5, pp. 2171–2174, 24th Apr. 2002, ISSN: 0034-6748. DOI: 10.1063/1.1464654. [Online]. Available: http://aip.scitation.org/doi/abs/10.1063/1.1464654 (visited on 17/10/2017).

[126] W. F. Walls and F. L. Walls, 'A practical guide to isolation amplifier selection', in *2007 IEEE International Frequency Control Symposium Joint with the 21st European Frequency and Time Forum*, May 2007, pp. 557–562. DOI: 10.1109/FREQ.2007.4319134.

[127] L. Sze-Ming, 'Influence of noise of common oscillator in dual-mixer time-difference measurement system', *IEEE Transactions on Instrumentation and Measurement*, vol. IM-35, no. 4, pp. 648–651, Dec. 1986, ISSN: 0018-9456. DOI: 10.1109/TIM.1986.6831788.

[128] F. L. Walls and F. G. Ascarrunz, 'The effect of harmonic distortion on phase errors in frequency distribution and synthesis', in *NIST*, 1st Jan. 1995. DOI: 105628. [Online]. Available: https://www.nist.gov/publications/effect-harmonic-distortion-phase-errors-frequency-distribution-and-synthesis (visited on 30/10/2017).

[129] F. L. Walls, S. R. Stein, J. E. Gray and D. J. Glaze, 'Design considerations in state-of-the-art signal processing and phase noise measurement systems', in *30th Annual Symposium on Frequency Control*, Jun. 1976, pp. 269–274. DOI: 10.1109/FREQ.1976.201325.

[130] J. Ackermann, Ed. (2017). The time-nuts archives, [Online]. Available: https://www.febo.com/pipermail/time-nuts/ (visited on 16/10/2017).

[131] B. Griffiths. (2017). Bruce's precision time and frequency circuits, notes, references, and tips, [Online]. Available: http://www.ko4bb.com/~bruce/index.html (visited on 16/10/2017).

[132] L. M. Nelson and F. L. Walls, 'Environmental effects in mixers and frequency distribution systems', in *Proceedings of the 1992 IEEE Frequency Control Symposium*, May 1992, pp. 831–837. DOI: 10.1109/FREQ.1992.269954.

[133] O. Collins, 'The design of low jitter hard limiters', *IEEE Transactions on Communications*, vol. 44, no. 5, pp. 601–608, May 1996, ISSN: 0090-6778. DOI: 10.1109/26.494304.

[134] U. Bangert. (2017). Ulrich bangert's homepage, Downloads, [Online]. Available: http://www.ulrich-bangert.de/html/downloads.html (visited on 06/11/2017).